Recent Topics In Semiconductor Physics

Symposium on

RECENT TOPICS IN SEMICONDUCTOR PHYSICS

In Commemoration of the Sixtieth Birthday of Professor Yasutada Uemura

29 March 1982, University of Tokyo, Japan

Editors

HIROSHI KAMIMURA
Department of Physics
University of Tokyo
Bunkyo-ku, Tokyo
Japan

YUTAKA TOYOZAWA
Institute for Solid State Physics
University of Tokyo
Roppongi, Minato-ku, Tokyo
Japan

World Scientific

World Scientific Publishing Co. Pte. Ltd.
P O Box 128
Farrer Road
Singapore 9128

ISBN 9971-950-54-5
9971-950-55-3 pbk

Printed in Singapore by Richard-Clay (S. E. Asia) Pte. Ltd.

PREFACE

This special issue entitled "Recent Topics in Semiconductor Physics" has been planned to commemorate the sixtieth birthday of Professor Yasutada Uemura, professor emeritus of the University of Tokyo, and currently a professor of physics at the Tokyo Science University.

Semiconductor physics is a major branch of physics. For the past thirty years Professor Uemura's academic activities have been highly appreciated in the field of semiconductor physics. In particular, as a theorist he has made major contributions to the development of the physics of the MOS electronic system. Thus, in order to commemorate his scientific achievement on the occasion of his retirement from the University of Tokyo, a special symposium was held on 29 March 1982. At the symposium talks were given on several topics of current interest which were close to the area of Prof. Uemura's research activities. This special volume is a collection of the review articles written by the seven speakers at that symposium.

It is our great pleasure to thank all the authors for their most valuable contributions to this special issue. Finally we would like to mention that the use of camera-ready manuscripts in the production of this volume was because all the topics covered here are of current interest and merit rapid publication.

July 1983

Hiroshi Kamimura
Yutaka Toyozawa

CONTENTS

A PERSPECTIVE IN SUPERLATTICE DEVELOPMENT

L. Esaki

IBM Thomas J. Watson Research Center
P. O. Box 218, Yorktown Heights, N.Y. 10598 USA

I. INTRODUCTION

Research on synthesized semiconductor superlattices was initiated with a proposal in 1969-1970 by Esaki and Tsu[1,2]: a one-dimensional periodic structure consisting of alternating ultra-thin layers, with its period less than the electron mean free path. Figure 1 illustrates a perfect superlattice structure prepared atomic layer by atomic layer with an ideal epitaxy.

The idea of the superlattice occurred to us when examining the possible observation of resonant tunneling through double and multiple barriers. Such resonant tunneling arises from the interaction of electron waves with potential barriers.[3] The DeBroglie wavelength λ is calculated to be 123Å for electrons with an energy E of 0.1eV and an effective mass m^* of $0.1m_o$ from the equation, $\lambda = h(2m^*E)^{-1/2}$, where h is Planck's constant. The electron mean free path, an important parameter for the observation of quantum effects, depends heavily on crystal quality and also on temperatures, the effective mass, etc. If characteristic dimensions such as a superlattice period, layer thicknesses (widths of potential wells or barriers), etc., in semiconductor nanostructures are reduced to less than the electron mean free path, the entire electron system enters into a quantum regime with the assumption of the presence of ideal interfaces, as schematically illustrated in Fig. 2. Our effort for the semiconductor superlattice is viewed as a search for novel phenomena in such a regime.

In this review, we first describe highlights in the early period (1969-1974) when most of the superlattice studies, including the structure synthesis by Molecular Beam Epitaxy (MBE), were being carried out in our group.[4] This pioneering work is believed to have provided the foundation for subsequent progress, which proceeded rather rapidly in scope as well as in depth, stimulating explorations on quantum wells and other heterostructures. Our early efforts focussed on transport properties in the direction of the one-dimensional periodic potential. Later, however, it was noticed that optical studies such as absorption, luminescence, lasers, magneto-absorption and inelastic light scattering, and magneto-quantum transport of the two-dimensional electron gas system in the layer plane prevailed. Such expanded activities resulted in new discoveries and inventions, leading to further proliferation of heterostructures in general. Some of the most significant developments in recent years will be summarized in Chapter III.

II. EARLY PERIOD

1) Proposal of a Semiconductor Superlattice

Esaki and Tsu[1,2] envisioned two types of synthesized superlattices: doping and compositional, as shown in the top and bottom of Fig. 3, respectively, where, in either case, a superlattice potential was introduced by a periodic variation of impurities or composition during epitaxial growth. The techniques of thin films, then, were rapidly advancing. It was theoretically shown that such a synthesized structure possesses unusual electronic properties not seen in the host semiconductors, arising from predetermined quantum states of two-dimensional character.

The introduction of the superlattice potential clearly perturbs the band structure of the host materials: The degree of such perturbation depends on its amplitude and periodicity. Since the superlattice period ℓ is usually much greater than the original lattice constant, the Brillouin zone is divided into a series of minizones, giving rise to narrow allowed subbands, E_1 and E_2, separated by forbidden regions in the conduction band of the host crystal, as shown at the top of Fig. 4. Thus, this results in a highly perturbed energy-wave vector relationship for conduction electrons, as schematically illustrated at the bottom of Fig. 4. Figure 5 shows the density of states $\rho(E)$ for electrons in a superlattice in the energy range including the first three subbands: E_1 between a and b, E_2 between c and d and E_3 between e and f (indicated by arrows in the figure), in comparison with the parabolic curve for the three-dimensional electron system and the staircase-like density of states for the two-dimensional system.

Although the situation is analogous to the Kronig-Penney band model,[5] it is seen here that the familiar properties are observed in a new domain of physical scale. In an extreme case, where quantum wells are sufficiently apart from each other, allowed bands become discrete states, and then electrons are completely two-dimensional in which the density of states is illustrated by the dashed line in Fig. 5.

The allowed subbands in a superlattice can be calculated from the following expression, assuming a one-dimensional, periodic square-well potential,[6]

$$-1 \leq \cos \frac{a(2mE)^{1/2}}{\hbar} \cosh \frac{b[2m(V-E)]^{1/2}}{\hbar} +$$

$$\left(\frac{V}{2E} - 1\right)\left(\frac{V}{E} - 1\right)^{-1/2} \sin \frac{a(2mE)^{1/2}}{\hbar} \sinh \frac{b[2m(V-E)]^{1/2}}{\hbar} \leq 1,$$

where E is the electron energy in the superlattice direction, V, the barrier height (0.4V), a, the well width, b, the barrier width, and m, the effective mass ($0.1m_o$). Assuming equal width (a=b), the allowed bands, E_1, E_2, E_3 and E_4 calculated as a function of the width are shown with the hatched regions in Fig. 6, including discrete energy levels for a single square well with the width a.

The dynamics of conduction electrons in such narrow bands was analyzed: A simplified path integration method[7] was used to obtain a relation between the applied field F and the average drift velocity v_d. The equations of motion are

$$\hbar \partial k_x/\partial t = eF \text{ and } v_x = \hbar^{-1} \partial E_x/\partial k_x;$$

the velocity increment in a time interval dt is

$$dv_x = eF\hbar^{-2}(\partial^2 E_x/\partial k_x^2)dt.$$

The average drift velocity, taking into account the scattering time τ, is written as

$$v_d = \int_0^\infty \exp(-t/\tau)dv_x$$

$$= eF\hbar^{-2}\int_o^\infty (\partial^2 E_x/\partial k_x^2) \exp(-t/\tau)dt$$

If a sinusoidal approximation is used for the E_x - k_x relationship,

$$v_d = \frac{eF\tau}{m^*(0)}\{1 + (\frac{eF\tau\ell}{\hbar})^2\}^{-1},$$

thus the current-field curve is given by

$$J = env_d,$$

where $m^*(o)$ is determined by the curvature of $E_x(k_x)$ at $k_x = o$ and n is the electron concentration. The current J, plotted as a function of F in Fig. 6, has the maximum at $eF\tau\ell/\hbar = 1$ (indicated by an arrow) and thereafter decreases, corresponding to a decreasing average drift velocity, giving rise to a differential negative resistance. This result indicates that, if, in applying modestly high electric fields, conduction electrons gain enough energies to reach beyond inflection points (indicated by arrows in Fig. 4), they will be decelerated rather than accelerated by such electric fields.

It is worthwhile mentioning that, in 1974, Gnutzmann et al.[8] pointed out an interesting possibility; namely, the occurrence of a direct-gap superlattice made of indirect-gap host materials, because of Brillouin-zone folding as a result of the introduction of the new superlattice periodicity, which was later reexamined by Madhukar.[9] The idea may suggest the synthesis of new optical materials.

2) Experimental Efforts

In 1970, Esaki et al.[10] reported the first experimental result on a superlattice synthesized by Blakeslee and Aliotta[11] with the CVD technique. The structure was obtained by a periodic variation of the phosphorus content, x, in the $GaAs_{1-x}P_x$ system. The period was made as thin as 200Å with x varying from 0-0.1 to 0.5. Although the superlattice formation was clearly demonstrated by electron microscopy, X-ray diffraction and cathode luminescence, transport measurements failed to show any predicted quantum effect. In this system, a relatively large lattice-mismatch, 1.8%, between GaAs and $GaAs_{0.5}P_{0.5}$ inevitably generates a strain at the interfaces.

It was recognized at the beginning that, while the structure was undoubtedly of considerable theoretical interest, the formation of such a refined layer-structure would be a formidable task. Nevertheless, the proposal of a semiconductor superlattice inspired a number of material scientists. In addition to the above-mentioned effort, Cho[12] and Woodall[13] demonstrated that it was feasible to produce alternating layers of lattice-matched GaAs and $Ga_{1-x}Al_xAs$ with the MBE and LPE techniques, respectively, while it is now well known that the MBE technique prevails. In 1972, we established a MBE system particularly designed for the superlattice structure, which led to the successful growth of a GaAs-GaAlAs superlattice.[14] Esaki et al.[15] found that this superlattice exhibited a negative resistance in its transport properties, which was, for the first time, interpreted on the basis of the predicted quantum effect. Figure 8 shows the current-voltage characteristic at room temperature for such a superlattice having 100 periods with each period consisting of a GaAs well 60Å thick and a $Ga_{0.5}Al_{0.5}As$ barrier 10Å thick. The structure was grown between two GaAs layers doped to an electron concentration of $5x10^{17}cm^{-3}$ on a similarly doped substrate. The electron concentration in the superlattice region is estimated to be approximately $10^{16}cm^{-3}$. The curve trace of Fig. 8, quite similar to Fig. 7 theoretically predicted, was obtained from a 100ns pulse measurement with a repetition rate of 10kHz. The samples were evaluated with various techniques including HEED, He-ion backscattering, and Raman scattering. The compositional profile of such a structure was directly verified by the simultaneous use of ion sputter-etching of the sample surface and Auger electron spectroscopy.[16]

3) Multibarrier Tunneling

In 1973, Tsu and Esaki[17] computed the transport properties from the tunneling point of view, leading to the derivation of I-V curves. The superlattice band model previously presented, assumes an infinite periodic structure. In reality, however, not only a finite number of periods is prepared with alternating epitaxy, but also the electron mean free path is limited. In addition, two terminal electrodes for transport measurements, which create interfaces between the superlattice and such electrodes, are unavoidable. The potential profile of the realistic system is schematically illustrated in Fig. 9, where the top and the bottom are an n-period superlattice of a finite length ℓ used for calculations of the reflection R and transmission T amplitudes, and the tunnel current J as a function of applied voltage V, respectively. Assuming the periodic structure made of GaAs and GaAlAs layers, the barrier height and width, and the well width are taken to be 0.5eV, 20Å and 50Å, respectively, for calculations. The electrodes are made of n-GaAs layers, of which the Fermi energy is indicated as E_f in the figure. Incidentally, one will notice in this system that one-dimensional localization results from the fact that the two terminal electrodes serve as reservoirs which randamize the phase of electron waves.

The three-dimensional Schrödinger equation for the one-dimensional period potential U(x), where x is the direction of the multibarrier, can be separated into transverse and longitudinal parts; namely, the total energy E is written by the sum of the longitudinal and transverse energies,

$$E = E_{\ell}(U) + \hbar^2 k_t^2/2m^*,$$

and the wave function is expressed by the product,

$$\psi = \psi_{\ell} \bullet \psi_t .$$

The electron wave functions in the left and right contacts are

$$\psi_1 = \psi_{\ell}[\exp(ik_1 x) + R\exp(-ik_1 x)],$$

$$\psi_n = \psi_t[T\exp(ik_n x)],$$

respectively. By matching wavefunctions and their first derivatives at each interface, one can derive the reflection and transmission amplitudes, R and T. In Fig. 10, the transmission coefficient T*T is plotted as a function of electron energy for a double barrier. As shown in the insert of the figure, if the energy of incident electrons coincides with the resonant bound states within the potential well, the electrons tunnel through both barriers without attenuation. Such unity transmissivity arises because, for the electron waves of the resonant energies E_1 and E_2, the reflected waves from inside interfere destructively with those from outside, so that only transmitted waves remain.[3] In Fig. 11, *ℓn* T*T is plotted as a function of electron energy for a double, a triple and a quintuple barrier structure. Note that the resonant energies for the triple-barrier (double-well) case are split into doublets, and those for the quintuple barrier (quadruple-well) case are split into quadruplets. In the double-well case, each single-well quantum state is split into a symmetric and an antisymmetric combination. This is extended to n wells. In the large n limit, a band-like condition is ultimately achieved, which corresponds to the superlattice band model.

Because of a separation of variable, the transmission coefficient is only a function of electron energy in the longitudinal direction. The tunnel current J, integrated over the transverse direction, is given by

$$J = \frac{em^*kT}{2\pi^2\hbar^3}\int_o^\infty T^*T\ell n\left(\frac{1+\exp[(E_f-E_\ell)/kT]}{1+\exp[(E_f-E_\ell-eV)/kT]}\right)dE_t$$

where E_ℓ is the electron energy in the longitudinal direction. In the low-temperature limit, the above expression becomes

$$J = \frac{em^*}{2\pi^2\hbar^3}\int_o^{E_f}(E_f-E_t)T^*TdE_\ell, V \geq E_f$$

and

$$J = \frac{em^*}{2\pi^2\hbar^3}\left[V\int_o^{E_f-V}T^*TdE_\ell + \int_{E_f-V}^{E_f}(E_f-E_t)T^*TdE_\ell\right], V < E_f.$$

We believe that this multibarrier tunneling model provided a useful insight into the transport mechanism and laid the foundation for the following experimental investigations.

4) Search for Quantum States

In early 1974, Chang et al.[18] observed resonant tunneling in double barriers, and subsequently Esaki et al.[19] measured quantum transport properties for a superlattice having a tight-binding potential.

The I-V and dI/dV-V characteristics are shown in Fig. 12 for a double barrier with a well of 50Å and two barriers of 80Å made of $Ga_{0.3}Al_{0.7}As$. The structure is completed by having two outside GaAs electrodes doped to

10^{18} cm^{-3}: the substrate on one side and a top overgrowth on the other. The schematic energy diagram is shown in the inset where the first two quantized levels are indicated.

The resonance, in this case, is achieved under applied voltages by aligning the Fermi level of the electrode with the bound quantum states. The current should show peaks at voltages equal to twice the energies of the bound quantum states referred to the original conduction band edge in the well. The situation is shown as (a) and (c) in the energy diagrams and is also marked correspondingly in the experimental curves in Fig. 12, where two sets of singularities are clearly visible. The current at the lower level shows only a slight hump, but that at the higher level shows a pronounced peak and thus a negative resistance. The situation at the opposite voltage polarity is similar except that the singularities occur at somewhat lower values; the asymmetry is believed to be due to nonidentical barriers or an asymmetric barrier shape. The broadness of the singularities comes mainly from the fluctuations in both thicknesses and potentials that may exist in the formation of the double-barrier structure. A large number of double barriers were made with three different well widths, 40, 50, and 65Å. Although there is some understandable spread, measured values for the quantized levels have been verified to coincide with the calculated ones. This agreement undoubtedly has confirmed the observation of resonant tunneling and, consequently, the energies of the bound quantum states.

Transport properties were investigated on a superlattice which comprises 50 periods, with each period consisting of a GaAs well 45Å thick and an AlAs well 40Å thick. The electron concentrations in the superlattice region and contacting GaAs regions are approximately 10^{17} and $10^{18}cm^{-3}$, respectively. Because of a tight-binding potential, the conduction bandwidth is estimated to be as narrow as 5meV in comparison with 40meV for the previous superlattice,[15] presented in Fig. 8. The I-V and dI/dV-V characteristics of this superlattice are shown in Fig. 13. The current at low voltages will be carried by a marginal band-conduction mechanism. As the voltage is increased to a value such that the drop across each period is comparable with the bandwidth, the band conduction cannot be maintained. This decrease of overlapping of the wavefunctions reduces the current, resulting in a negative differential resistance. With further increase in applied voltage, electrons will proceed by tunneling across a newly created, localized high field domain. Slight nonuniformity in the shape of barriers or wells will lead to domain formation, which is indeed inherent in media having a voltage-controlled negative resistance. Once the first domain is formed, the tunneling process across it limits the current and keeps band-type conduction alive in other regions. This situation is maintained until the first band and the second band across this domain become aligned. A current maximum then occurs and the domain expands. The cases marked by (b) and (c) in Fig. 13 illustrate the events. The conductance in the negative resistance region is complicated by the association of instabilities. At higher applied voltages, however, a well-behaved oscillatory phenomenon is seen with a period of 0.24V. As expected, this value agrees with the energy difference between the first and second allowed bands, calculated to be 0.24eV for the present superlattice.

These experiments probably constitute the first clear observation of quantam states in both single and multiple potential wells. The impact of the achievement of superlattices or multibarrier strucutres is many-fold. The elegance of one-dimensional quantum physics, which had long remained a textbook exercise, could now, for the first time, be practiced in a laboratory (Do-It-Yourself Quantum Mechanics!). It should be emphasized that such an achievement would not have been accomplised without the key contribution of material preparation techniques.

During the last ten years, the MBE technique was further perfected, producing smooth layers and abrupt changes in ultrathin dimensions. It is not surprising that, with such an improvement in technique, Sollner et al.[20] recently demonstrated a dramatical improvement in the current singularity and negative resistance in resonant double-barrier tunneling through a GaAs quantum well and carried out detecting and mixing at frequencies as high as 2.5 THz.

5) X-ray Analysis

For analyses of GaAs-$GaAs_{1-x}P_x$[10] and GaAs-AlAs[21] superlattice structures, the high-angle as well as the low-angle X-ray scattering data provided relevant information[22] including the accurate determination of the superlattice period. Figure 14 shows the high-angle diffraction patterns for an 88Å-period GaAs-AlAs superlattice, obtained in the vicinity of (a) the 002, (b) the 004, and (c) the 006 reflections. The substrate peak is the highest one in the center. To its left in (b) and (c), the zero-order reflection of the superlattice is visible. The weak superlattice reflections are designated as +1, +2, +3, and -1, -2, -3 satellites. Figure 15 shows low-angle X-ray interferences for a 90.5Å-period superlattice, where several interference maxima are observed corresponding to the superlattice period. From this data, the numbers of Ga and Al layers in the superlattice unit cell can be accurately determined. On the other hand, the relative intensities of the high-angle reflections (Fig. 14) are a sensitive measure of the elastic strain present in the lattice. The measured elastic strain agreed with the value computed theoretically on the assumption that the strain is not relieved by misfit dislocations at the GaAs-AlAs interfaces.

In any case, some deviations from the ideal structure are evident from the fact that the half-widths of the superlattice reflections vary and the fit between the observed and calculated satellite intensities is not exact. At the GaAs-AlAs interface there will be mixed layers even if there is no diffusion because at the instant one beam is shut off and the other beam switched on, the surface will, in general, have a partial layer with an atomic step in it. This causes a rounding of the edges of the composition profile at the interface and produces incommensurate superlattice periods. Recently, Weisbuch et al.[23] reported some estimates of interface roughness for GaAs-GaAlAs multiquantum well structures, analyzed from broadening of luminescence linewidths, which suggested the presence of the island-like structure with a one monolayer height and a lateral size of 300Å or more.

6) Comments on Interfaces

One should realize that the superlattice proposal by Esaki and Tsu[1,2] was made on a rather optimistic presupposition of a "clean" and yet atomically smooth interface which provides only an abrupt potential step with little undesirable localized states. Recently, considerable efforts have been made to understand the electronic structure at interfaces[24] or heterojunctions.[25] Even in an ideal situation, the discontinuity at the interface provides formidable tasks in theoretical handling: Propagating and evanescent Bloch waves should be matched across the interface, satisfying continuity conditions on the envelope wave functions.[26,27] The interface chemical bonds differing from the bulk may give rise to localized states in addition to bond-relaxation in some circumstances.

In reality, however, the presence of possible misfit dislocations and other defect-complexes, as well as roughness or disorder at the interface may not be completely avoidable even with most advanced techniques. Indeed, the early experimental results described in this chapter suffered considerably from such departures from an idealized model on which simple theories are based. The extrinsic effects due to localized states around defect sites near the interface should be distinguished from the intrinsic properties arising from the discontinuities of the wave functions. In favorable systems such as GaAs-GaAlAs, the intrinsic localized states are virtually nonexistent in the energy range in which we are interested.

III. IMPORTANT DEVELOPMENTS

1) Remarks

Following the pioneering period, a series of interesting developments have been witnessed on this subject. Expanding activities resulted in observations of new phenomena and exploration into novel structures, leading to ramification of the superlattice field. Such structures include quantum-wells and nearly ideal heterojunctions.

In this Chapter, we will try to describe some of the important developments in chronological order, as much as possible, by categorizing them according to host materials used, GaAs-$Ga_{1-x}Al_xAs$ and InAs-GaSb(-AlSb), followed by all other systems. Such categorization, however, often encounters difficulties because similar studies extend over a variety of materials; the same experimental techniques and theoretical methods can be easily applied to different systems. Nevertheless, most significant studies from scientific as well as technological aspects have been carried out with GaAs-$Ga_{1-x}Al_xAs$ and InAs-GaSb systems. In Fig. 16, three typical examples for the relationship of bandedge energies at heterojunctions (left) and their respective superlattices or multiquantum-wells (right) are shown together with corresponding band-bending and carrier confinement at the interface (middle). The conduction band discontinuity ΔE_c is equal to the difference in the electron affinities of the two semiconductors. In case (c), the band-bending and electron and hole confinements occur, as indicated, regardless of the polarity of employed semiconductors, whereas, in cases (a) and (b), those depend on the semiconductor polarity (N or P).

As seen in Fig. 16, in the GaAs-$Ga_{1-x}As_xAs$ system, the total difference in the energy gaps is shared by potential steps ΔE_c and ΔE_v, while the InAs-GaSb system exhibits a rather unusual bandedge relationship; namely, the conduction bandedge of InAs, E_{c1}, is lower than the valence bandedge of GaSb, E_{v2}, by E_s. The superlattices made of these three types of semiconductor pairs are identified as type I, type I' and type II, as illustrated from the top to the bottom on the right side. Although a variety of superlattices have been synthesized with many semiconductor pairs, as listed in Table I, all of them will fall into these three categories.

2) GaAs-GaAlAs

i) Optical Absorption

Dingle et al. observed pronounced structure in the optical absorption spectrum, representing bound states in isolated[28] and coupled quantum wells.[29] For the former, GaAs well widths in the range between 70Å and 500Å were prepared. The GaAs wells were separated by $Ga_{1-x}Al_xAs$ barriers which were normally thicker than 250Å. In low-temperature measurements for such structures, several exciton peaks, associated with different bound-electron and bound-hole states, were resolved. For the latter study, a series of structures, with GaAs well widths in the range between 50Å and 200Å and $Ga_{1-x}Al_xAs$ $(0.19<x<0.27)$ barrier widths between 12Å and 18Å, were grown by MBE on GaAs substrates. The spectra at low temperatures clearly indicated the evolution of resonantly split discrete states into the lowest band of a superlattice. In all experiments, in order to enhance the total GaAs absorption, as many as fifty (or eighty) GaAs layers were grown in a single structure. These observations demonstrated the great precision of MBE in fabricating thin and uniform layers.

ii) Lasers

van der Ziel et al.[30] observed optically pumped laser oscillation from the above-mentioned quantum-well structures at 15K. The laser oscillation occurred at energies which are slightly below the exciton associated with the lowest energy $n = 1$ bound state in GaAs quantum wells (50-500Å thick).

In 1978, Dupuis et al.[31] and Holonyak et al.[32] succeeded in room-temperature laser operation of quantum-well $Ga_{1-x}Al_xAs$-GaAs laser diodes, where the well width was around 200Å. They observed lasing transitions all the way to the $n=5$ bound state in the quantum well corresponding to 1.8eV (0.69 micron), which is 0.29eV higher than that of bulk GaAs. It should be added that these devices were prepared by metalorganic chemical vapor deposition (MOCVD). In view of the recent progress of the MOCVD, this technique appears to be a serious contender to the MBE technique for the growth of quantum-wells or superlattices.

More recently, Tsang[33] succeeded in attaining a threshold current density J_{th} as low as 250A/cm^2 in MBE-grown $Ga_{1-x}Al_xAs$-GaAs lasers with a multiquantum-well structure as shown in Fig. 17. This was achieved as a result of utilizing the beneficial effects arising from the two-dimensional density of states of the confined carriers (Fig. 5). It is generally observed that, in mutliquantum-well lasers, the beam width in the direction perpendicular to the junction plane and the temperature dependence of J_{th} are significantly reduced in comparison with the regular double-heterostructure (DH) lasers.

iii) Photocurrent and Luminescence

Tsu et al.[34] made photocurrent measurements which shed light on the interrelationship between quantum states and anomalous transport properties in a superlattice. Superlattice structures of three different configurations, designated A, B and C, were grown on n-type 10^{18} cm^{-3} GaAs substrates: A, 100 periods of 35Å GaAs-35Å $Ga_{0.8}Al_{0.2}As$; B, 80 periods of 50Å GaAs-50Å $Ga_{0.78}Al_{0.22}As$; C, 50 periods of 110Å GaAs-110Å $Ga_{0.55}Al_{0.45}As$. The total thickness in all cases is of the order of one micrometer, which is comparable with the absorption length for the photons involved. The superlattice region is undoped and an ohmic contact is made on the bottom of the substrate. A semi-transparent Au film about 100Å thick serves as the top electrode.

The energy diagram is shown schematically in the upper part of Fig. 18, where quantum states created by the periodic potential are denoted E_1 and E_2. These states are essentially discrete if there is a relatively large separation between wells, as is the case in sample C, but broaden into bands in samples A and B owing to an increase in overlapping of wavefunctions.

Shown in Fig.18 are the normalized photocurrent versus photon energy curves, measured at 5°K under zero bias. As the period becomes narrower, the photocurrent increases and the peak positions shift to higher energies with fewer numbers being observed. In particular, because of the discrete nature of states in sample C, the observed photocurrent is much smaller than those in samples A and B. Calculated energies are indicated by arrows, where E_n is the transition from the nth quantum state of heavy holes in the valence band to that in the conduction band. The calculated combined bandwidth for each transition in samples A and B is also shown, which is mainly determined by electrons.

More recently, Mendez et al.,[35] with a somewhat similar experimental arrangement, observed an electric field-induced quenching of photoluminescence from quantum wells. Six identical GaAs quantum wells (width = 35Å) in GaAlAs medium were formed in the space charge region of the Schottky barrier, as schematically illustrated in Fig. 19. With an increasing applied field perpendicular to the quantum wells, the luminescence intensity decreases and becomes completely quenched at an average field of a few tens of kV/cm, (applied voltage = -0.4V), as shown in Fig. 20. This is accompanied by a shift to lower energies of the peak positions. The observation is interpreted as caused by the field, which induces a separation of electrons and holes in the quantum well with the modification of the quantum states in the wells, resulting in luminescence quenching.

iv) Raman Scattering

Manuel et al.[36] reported the observation of enhancement in the Raman cross section for photon energies near electronic resonance in GaAs-$Ga_{1-x}Al_xAs$ superlattices of a variety of configurations. Both the energy positions and the general shape of the resonant curves agree with those derived theoretically based on the two-dimensionality of the quantum states in such superlattices. Polarization studies indicate a major contribution to the scattering from forbidden processes.

Among recent developments in the field have been the introduction of inelastic light scattering as a spectroscopic tool in the investigation of the two-dimensional electron system. The significance of resonant inelastic light scattering was first pointed out by Burstein et al.,[37] claiming that the method yields separate spectra of single particle and collective excitations which will lead to the determination of electronic energy levels in quantum wells as well as Coulomb interactions. Subsequently, Abstreiter et al.[38] and Pinczuk et al.[39] observed light scattering by intersubband single particle excitations, between discrete energy levels, of two-dimensional electrons in GaAs-$Ga_{1-x}Al_xAs$ heterojunctions and quantum wells.

Meanwhile, Colvard et al.[40] reported the observation of Raman scattering from folded acoustic longitudinal phonons in a GaAs(13.6Å)-AlAl(11.4Å) superlattice. The superlattice periodicity is expected to result in Brillouin-zone folding (as previously mentioned) and the appearance of gaps in the phonon spectrum for wave vectors satisfying the Bragg condition. An explanation of the data was given with a simple theory which involves the Kronig-Penney electron model and the phonons in the elastic continuum limit.

v) Magnetoquantum Effects

In 1977, Chang et al.[41] reported the first observation of the oscillatory magnetoresistance (Shubnikov-de Haas effect) in GaAs-$Ga_{1-x}Al_xAs$ superlattices with the current flowing in the plane of the layers. The superlattice provides a unique opportunity to create made-to-order Fermi surfaces by controlling the energy and bandwidth of the subbands which are determined by the barrier and well-thicknesses as well as the barrier height. The observed oscillations manifest the electronic subband structure, which becomes increasingly two-dimensional in character as the bandwidth is narrowed.

Figure 21 illustrates Fermi surfaces in the k_t - k_z plane and the densities of states (t refers to the x-y plane - the plane of the layers) calculated for three superlattice specimens: A, 40Å GaAs - 30 Å $Ga_{0.84}Al_{0.16}As$, $n = 1.4 \times 10^{18}$ cm^{-3}; B, 90 Å GaAs - 75Å $Ga_{0.83}Al_{0.17}As$, $n = 1.2 \times 10^{18} cm^{-3}$; and C, 90Å GaAs - 90 Å $Ga_{0.89}Al_{0.11}As$, $n = 1.9 \times 10^{18}$ cm^{-3}, with a total thickness of about 2 micrometers, where n is the electron concentration. In specimen A, a relatively weak potential gives a rather broad subband and the Fermi surface only slightly deviated from a sphere. An increase in the superlattice potential enhances this anistropy, leading toward two-dimensionality; namely, the energy-independent density of states and the cylindrical Fermi surface. Specimen B represents nearly such a case. The situation in specimen C can be looked upon as a composite case in that electrons occupy both the narrow ground subband as well as the relatively broad second subband.

The percentage change of the magnetoresistance at 4.2K as a function of magnetic field is shown in Fig. 22 for specimen C. Long-bar-shaped geometry, as illustrated in the inset, was used throughout the present experiment. The current flows in the x direction while the magnetic field is applied in the y-z plane with the angle ψ defined with respect to the z axis. It is seen that, in the region of high angles, well-defined oscillations exist with the extremal points apparently shifting toward higher fields and gradually disappearing as ψ is increased. Their evanescence at high angles reveals that there remains a different, albeit less pronounced, set of oscillations. Now, the former well-defined oscillations at low angles are attributed to the ground subband of the two-dimensional character, whereas the latter less pronounced oscillations at high angles are likely associated with electrons in the second subband, being somewhat three-dimensional in character, could cause oscillations even in this field orientation. The detailed analysis of the observed oscillatory behaviors indicated a good agreement with the Fermi surfaces calculated on the basis of the superlattice configuration and the electron concentration.

In 1980, Tsui et al.[42] reported the first observation of magnetophonon resonances in the two-dimensional electron system in GaAs-GaAlAs heterojunctions and superlattices. The resonances, detected in the magnetic field dependence of the samples resistance at temperatures between 100 and 180K, result from inelastic scattering of electrons between Landau levels with resonant absorption of GaAs LO phonons under the resonance condition of $N\hbar\omega_c = \hbar\omega_0$ where $\hbar\omega_c$ is the Landau level spacing, ω_0, the phonon frequency and N, an integer. The result showed that, in this system, the electron-phonon interaction with the polar mode of GaAs is dominant and its strength is consistent with that expected from polar coupling of two-dimensional electrons to the LO phonons of bulk GaAs. Englert et al.[43] made more detailed studies on this effect in higher magnetic fields so that the fundamental resonance (N=O) can be observed.

Though the two-dimensional electron systems, including the Si surface inversion layer, have been intensively studied under high magnetic fields for almost two decades, the experimental methods customarily employed are magnetoresistance and Hall effect. In both cases, however, the step from the experimental data to the density of states is difficult because electronic transport in a combination of the density of states and electron scattering. In 1983, Störmer et al.[44] observed the oscillatory magnetic moment as a function of magnetic field (de Haas-van Alphen effect) at 1.5K for two-dimensional electrons in a modulation-doped GaAs-GaAlAs heterostructure. Although the strength of the signal was enhanced by stacking 4000 layers, the amplitude of the oscillations was about a factor of 30 weaker than expected, which was attributed to sample inhomogeneities. Nevertheless, the measurements of the low temperature magnetic moment and susceptibility provide a useful tool for the investigation of the density of states per se.

vi) Modulation Doping and High Speed Devices

It is usually the case that free carriers, electrons or holes, in semiconductors are created by doping impurities: donors or acceptors. Thus, carriers inevitably suffer from impurity scattering. There are a few exceptions. Insulated-gate effect devices (MOSFET) are one such example, where electrons or holes are induced by applied gate voltages. InAs-GaSb superlattices may be another example where electrons and holes are produced solely by the unique bandedge relationship, as described later.

Now, in superlattices, it is possible to spatially separate carriers and their parent impurity atoms by doping impurities in the regions of the potential hills, as shown in Fig. 23. In the original article,[1] this concept was expressed in the general terms as follows, "....the scattering time is an important factor in the described effect,...., if the superlattice structure is formed in such a manner that most scattering centers such as foreign atoms, imperfections, etc., are concentrated in the neighborhood of the potential hills, one can show that electrons would suffer less from such scattering center..."

In 1978, Dingle et al.[45] successfully implemented such a concept in modulation-doped GaAs-GaAlAs superlattices, as illustrated in the top of Fig. 23, achieving electron mobilities which exceed the Brooks-Herring predictions. Modulation doping was performed by synchronizing the silicon (n-dopant) and aluminum fluxes in the MBE, so that the dopant was distributed only in the GaAlAs layers and was absent from the GaAs layer.

In 1980, Mimura et al.[46] and Delagebeaudeuf et al.[47] applied modulation-doped GaAs-GaAlAs heterojunctions to fabricate a new high-speed FET called HEMT (High Electron Mobility Transistor) or TEGFET (Two-Dimensional Electron Gas Field Effect Transistor), of which the energy diagram is shown in the bottom of Fig. 23. The device, if operated at 77°K, apparently exhibits a high-speed performance three times superior to that of the conventional GaAs MESFET.

vii) Quantized Hall Effect

In 1980, Klitzing et al.[48] demonstrated an interesting proposition that quantized Hall resistance can be used for precision determination of the find structure constant α, using two-dimensional electrons in the inversion layer of a Si MOSFET (Metal-Oxide-Semiconductor Field Effect Transistor). Subsequently, Tsui et al.[49] found that, with the modulation-doped GaAs-GaAlAs heterostructures, the pronounced characteristic of the quantized Hall effect can be easily observed, primarily because of their high electron mobilities, which led to the determination of α with a great accuracy such as α^{-1} = 137.035965(12)(0.089ppm).[50]

The quantized Hall effect in the two-dimensional electron system is observable at sufficiently high magnetic fields and low temperatures when there is little overlap in the density of state of neighboring Landau levels; in such a range of magnetic fields as to locate the Fermi level between the Landau levels, the parallel component of resistance ρ_{xx} vanishes and the Hall resistance ρ_{xy} goes through plateaus, as shown in Fig. 24. This surprising result can be understood by the argument that the localized states do not take part in quantum transport.[51] At the plateaus, the Hall resistance is given by $\rho_{xy} = h/e^2 i = \mu_o c/2\alpha i \approx 25813\ \Omega/i$ where i is the number of filled Landau levels, h the Planck's constance, e the electronic charge, μ_o the vacuum permeability, and c the speed of light in vacuum.

Recently, Tsui et al.[52] discovered a striking phenomenon: the existence of an anomalous quantized Hall effect, a Hall plateau in ρ_{xy} and a dip in ρ_{xx}, at a fractional filling factor of 1/3, as seen in Fig. 24, in the extreme quantum limit at temperatures lower than 4.2K. The result shown in Fig. 24 was recently obtained with a very dilute two-dimensional electron gas ($n = 6 \times 10^{10} cm^{-2}$) in the modulation-doped GaAs-GaAlAs heterostructures by Mendez et al.[53] Notice that the quantum limit (i=1) and a clear minimum corresponding to $\nu=1/3$ are reached at as low as 2.55T and 7.7T, respectively.

Laughlin,[54] as an explanation of such a fractional filling, presented variational ground-state and excited-state wave functions which describe the condensation of a two-dimensional electron gas into a new state of matter, an incompressible quantum fluid. The elementary excitations of this quantum fluid are fractionally charged, and this elegant theory predicts a series of ground states characterized by the variational parameter m (m=3, 5,...), decreasing in density and terminating in a Wigner crystal. More recently Störmer et al.[55] in addition to previously reported structures at filling factors, $\nu=1/3$ and 2/3, observed new structures at $\nu=4/3$, 5/3, 2/5, 3/5, 4/5, and 2/7. The results suggested that fractional quantization of the Hall effect exists in multiple series, each based on the inverse of an odd integer. There exists no full explanation for all of these experimental results at this moment.

viii) Impurity States and Excitons in Quantum Wells

Bastard[56] performed a variational calculation of the hydrogenic impurity ground states in a quantum well, obtaining the binding energy E as a function of well thickness and of impurity position. Figure 25(a) shows the binding energies in units of R^*_o: the three-dimensional effective Rydberg, at the edge and the center, indicated by crosses and dots, respectively, as a function of the normalized well thickness: L/a^*_o, where a^*_o is the three-dimensional effective Bohr radius. The electron confinement in a potential well lifts the usual ground bround state degeneracy with respect to the impurity position. The dependence of the binding energy on the impurity position leads to the formation of an impurity band. The density of states manifests itself in acceptor → conduction or valence → donor absorption process as two peaks located at the band extrema, as shown in Fig. 25(b), in which the dimensionless frequency χ is given by $\hbar\omega = E_g - E_i^{max} + \chi E_i^{max}$ where E_i refers to donor or acceptor binding energy. Photoluminescence from GaAs quantum wells measured by Miller et al.[57] apparently confirmed Bastard's prediction. Bastard et al.[58] derived the exciton binding energy in quantum wells with variation calculations. Similar calculations were reported by Green et al.[59]

ix) Excitonic Optical Bistability

In 1982, Gibbs et al.[60] reported excitonic optical bistability observed with a GaAs-GaAlAs superlattice etalon. In this case, laser light at frequencies just below the exciton absorption energy is slightly absorbed, producing carriers which change the absorptivity and the polarizability associated with the exciton resonance, thus giving rise to an intensity-dependent refractive index. The two-dimensionality of electrons confined in the quantum wells increases the binding energy of the free excitons in GaAs, permitting room-temperature bistable operation of such a superlattice etalon. This may suggest the possibility of a new optical bistable device.

x) Avalanche Photodiodes with Heterostructures

Superlattices or heterostructures have been exploited to obtain improved avalanche photodiodes. As is known, a large difference in the ionization rates for electrons and holes, α and β, respectively, is desirable for a low noise performance. Unfortunately, most III-V semiconductors have α nearly equal to β Therefore, Chin et al.[61] proposed increasing α/β by using a GaAs-GaAlAs superlattice structure, as shown in Fig. 26. In this structure, since the conduction bandedge discontinuity ΔE_c is considerably large than the valence bandedge discontinuity ΔE_v, the ionization rate for electrons will be enhanced, but not for holes, thus resulting in an artificial increase of α/β. Capasso et al.[62], indeed, obtained an increased α/β of 10 with a p-i-n structure consisting of 50 alternating 450Å GaAs and 550Å $Ga_{0.55}Al_{0.45}As$ layers. In addition, with the application of band-gap engineering such as graded gap or periodic structures, a number of new photodiodes or photomultipliers have been explored,[63] including channeling avalenche detectors[64] which spatially separate electrons and holes.

xi) Attempts for One-Dimensionality

The introduction of a superlattice potential facilitates the reduction of the dimensionality of carriers from three to two at the lower limit in the case of a series of quantum wells. On the other hand, the electrons on the surface inversion layers, as seen in Si MOSFETs, always constitute a two-dimensional gas. Thus, the electron system on the surface inversion layer of a superlattice crystal, as schematically illustrated in Fig. 27, will exhibit the dimensionality between one and two, depending upon the superlattice potential profile. It is expected here that, using a superlattice being p-type or having no carriers, surface electrons are generated either by an external field or by modulation-doping from a cladding GaAlAs layer (not shown in the figure). If the GaAlAs barriers in the superlattice are sufficiently thick so that wave functions in the electron pockets on the surface have no overlap, then a confined one-dimensional electron gas will be obtained.

Sakaki[65] proposed a V-grooved MOSFET of a one-dimensional electron gas with a single quantum well, virtually similar to that shown in Figure 27, suggesting extremely high electron mobilities on the basis of the scattering probability from Coulomb potential. Petroff et al.[66] attempted the realization of a one-dimensional carrier confining structure with MBE-grown GaAs and GaAlAs layers.

3) InAs-GaSb(-AlSb)

i) Theoretical Treatments

In 1976, in search for an alternative to the GaAs-AlAs system, in which the introduction of the superlattice periodicity will give a greater modification to the electronic band structure of the host materials, the InAs-GaSb system appeared as a good candidate because of its unique bandedge relationship at the interface, as shown in Fig. 16. At that time, of course, such alignment of the bandgaps of the two hosts was not well known. The bandedge relationship was estimated from the values of the electron affinities determined by Gobeli and Allen,[67] which might have had experimental uncertainties. Later, however, Sakaki et al.[68] confirmed that this was indeed the case, observing an unusual nonrectifying characteristic in the p-n junctions made of n-InAs and p-GaSb, which was the direct consequence of "interpenetration" between the GaSb valence band and the InAs conduction band. At the interface, shown in the center at the bottom of Fig. 16, electrons which "flood" from the GaAs valence band to the InAs conduction band, leaving holes behind, produce a strong dipole layer consisting of the two-dimensional electron and hole gases. An approximate classical solution for the band bending[69] can be obtained rather simply using a self-consistent Fermi-Thomas approximation.

First, Sai-Halasz et al.[70] made a one-dimensional calculation for the InAs-GaSb superlattices labeled type II at the bottom right-hand side of Fig. 16, treating each host material in Kane's two-band framework. The GaAs-AlAs superlattices, on the other hand, can be adequately treat in one-band approaches. In the InAs-GaSb superlattices, although potential wells for electrons and holes are located in the different semiconductors, quantized levels in the electron wells will be very close in energy to those in the hole wells. Thus, in such cases, there exists a strong interaction between them.

Following an analytical treatment, one obtains

$$\cos(kd_o) = \cos(k_1d_1)\cos(k_2d_2) - F\sin(k_1d_1)\sin(k_2d_2), \tag{1}$$

$$F = 1/2\left(\frac{ik_1 + u'_1/u_1}{ik_2 + u'_2/u_2} + \frac{ik_2 + u'_2/u_2}{ik_1 + u'_1/u_2}\right), \tag{2}$$

where the superlattice period d_o is the sum of the thicknesses d_1 and d_2 of alternating layers and $u_i = u_i(k_i,0)$, $u'_i = du_i(k_i,x)/dx\,|_{x=0}$. The allowed bands in the superlattice correspond to the energy range where Eq. 1 has a solution for real k. For such allowed bands, Eq. 1 provides the E-k relationship. In this calculation, we found that the energy gap is strongly dependent on the layer thickness or the period, decreasing from ~ 0.6eV to zero. It should be mentioned that the present result is reduced to the well-known Kronig-Penney solution, if the logarithmic derivatives u'_i/u_i in Eq. 2 are set to zero.

Next a band calculation was performed with the linear-combination-atomic-orbitals (LCAO) method[69] handling a large size of the primitive cell and ignoring charge redistribution at the interface. The calculated subband structure is strongly dependent upon the period. Figure 28 shows a potential profile and determined subband energies: E_i for electrons, LH_i and HH_i for light and heavy holes, respectively, for an InAs-GaSb superlattice with a period of 100Å, where E_{v2} - E_{c1} is assumed to be 0.15eV. Figure 29 shows calculated subband energies and bandwidths for electrons and light and heavy holes as a function of period, together with the energy gaps of GaSb and InAs in the left, assuming $d_1 = d_2$, where the semiconducting energy gap is determined by the difference E_1-HH_1. This gap decreases with increase in the period, becoming zero at 170Å, as seen in the figure, corresponding to a semiconductor-to-semimetal transition.

This indicates that, by a choice of the period, it is possible to synthesize a tailor-made narrow gap semiconductor or a semimetal with an electronic band structure which bears little resemblance to that of the host semiconductors. Once the system becomes semimetallic with increase in the period, the charge redistribution at the interface will give rise to severe band-bending.

ii) Optical Absorption and Luminescence

Optical absorption measurements[71] were made on InAs-GaSb superlattices with periods ranging between 30 and 60Å which are in the semiconductor regime as indicated in Figs. 28 and 29.

Samples consist of several hundred layers, typically 2.5-3μm thick, grown on GaSb substrates. The experiments were carried out by attaching the samples to a copper cryotip nominally at 10K. The samples were lapped to a thickness of about 50μm and polished from the bottom side of the substrates, on which monochromatic light was shone. There was no need to remove the substrate, for it is essentially transparent over the energy range of interest. The optical density of the superlattice was measured by registering the transmitted light intensity and comparing it with that similarly obtained from a bare substrate. The absorption coefficient was subsequently deduced, knowing the thickness of the samples.

For optical transitions, it should be noticed that the superlattice envelope wavefunctions are strongly dependent upon wavevectors. The matrix elements for subbands with low indices are quite small because the envelope wavefunctions of the conduction and valence subbands are mismatched, i.e., peaked in spacially different regions (electrons and holes are mostly confined in InAs and GaSb, respectively).

From measurements of the absorption coefficient α as a function of the photon energy, it is seen that the rise in α is rather weak in comparison with what it would ordinarily be at the bandedge of normal semiconductors, as theoretically expected. Furthermore, the absorption curves usually show fine structure, indicating the onset of transition between particular subbands. The measured absorption edges, E_{gs}, ranging from 0.25 to 0.37eV, are plotted as a function of the half period ($d_1=d_2$) as shown in Fig. 30.

The solid and broken lines in Fig. 30 are calculated superlattice energy-gaps for three different values of E_s: -100, -150, and -200 meV, where E_s is the energy difference between the top of the GaSb valence band, $E_v(B)$, and the bottom of the InAs conduction band, $E_c(C)$, as shown in Fig. 16. Although some uncertainties exist in the thickness measurements as well as the absorption-edge determination, it is clear that E_s falls in the vicinity of -0.15eV.

Recently, Voisin et al.[72] observed luminescence from the similar semiconducting InAs-GaSb superlattices. It was found that, in addition to radiative recombination between the electron and hole ground subbands, the luminescence spectra exhibit a low-energy tail which is believed to arise from impurities and interface defects.

iii) Semiconductor-Semimetal Transitions and Shubnikov-de Haas Oscillations

In order to verify the critical layer thickness[73] for the predicted semiconductor-semimetal transition, a number of InAs-GaSb superlattices with a variety of periods were grown on (100)GaSb substrates whose carriers largely freeze out at low temperatures. The total thickness of the superlattice region is typically ~2μm. No intentional doping was introduced, but residual impurities usually provide a background electron concentration of the order of $10^{16}cm^{-3}$. The approximate electron concentration and mobility parallel to the plane were derived from Hall measurements at 4.2K in low magnetic fields, because Hall voltages are primarily induced by electrons which have higher mobilities than holes.

The measured electron concentrations were plotted as a function of InAs layer thickness, as shown in Fig. 31. The thickness of the GaSb layer ranging from one-half of the InAs layer to about equal to that, is of secondary importance, since the energy gap is mainly determined by the ground subband in the conduction band of InAs.

It is evident from Fig. 31 that the electron concentration exhibits a sudden increase of an order-of-magnitude in the neighborhood of 100Å. This increase indicates the onset of electron transfer from GaSb to InAs: The transition from the semiconducting state to the semimetallic state, as the ground conduction subband, E_1, moves below the ground valence subband, H_1. The observation is in good agreement with the theoretical prediction, presented in Fig. 29, giving a critical layer thickness for this transition of 85Å, when the energy difference between the GaSb valence bandedge and the InAs conduction bandedge is 0.15eV at the interface.

It should be mentioned that, since this increase in the electron concentration is not due to doping of impurities, mobilities beyond 10^4cm/V•s at 4.2K can be achieved, much higher than bulk InAs with comparable electron concentrations, somehow similar to the modulation-doping.

The subband energy diagrams of two semimetallic superlattices: thin (top) and thick (bottom) are illustrated in Fig. 32, where it is clear that, as the layer thickness is widened, the band bending increases and more than one subband may become involved. As seen in Fig. 31, after reaching a peak, the carrier concentration decreases with increase in thickness and becomes saturated as the superlattice approaches the limit whereby it can be considered as a series of isolated heterojunctions. In this limit, most electronic conduction is taking place at heterojunction interfaces where electrons and holes are accumulated in a two-dimensional fashion.

Chang et al.[74] made extensive Shubnikov-de Haas experiments on superlattice samples covering the entire semimetallic regime from the onset of the semiconductor-semimetal transition to the heterojunction limit. Derivative techniques were used to enhance the oscillatory characteristics. The results of dR/dB up to a field of 10T are shown in Fig. 33 for three sample configurations: two samples with layer thicknesses of 200-100Å, and 300-150Å and one with that of 1000-1000Å, which is expected to be close to the heterojunction limit. Throughout this work the first thickness value is used to denote the InAs layer, and the second, the GaSb layer. Oscillations are observed in all cases and their amplitudes, generally speaking, tend to become increasingly pronounced with the layer thickness. For the 200-100Å sample, for instance, the second derivative has to be taken to show the well-defined oscillations in the low-field region. The oscillations disappear as the magnetic field is applied parallel to the plane of the layers, demonstrating the quasi-two-dimensional character of the semimetallic superlattices.

For samples with thin layers, such as that of 200-100Å, only the ground subband E_1 is occupied, as can be seen at the top of Fig. 32. The corresponding characteristic at the top of Fig. 33 exhibits clearly a single series of oscillations. The standard plot of the inverse fields of extrema positions vs. integers yields a straight line; from its slope the electron density and the Fermi energy can be derived.

Unlike the situation in the semiconducting state, semimetallic superlattices with high carrier mobilities exhibit large transverse magnetoresistance.[75] The ratio $R_{\perp}(B)/R_o$ vs. B is shown in Fig. 34 for two samples with the layer thicknesses of 120-80Å and 200-100Å, where $R_{\perp}(B)$ and R_o are the resistances at B and at B=0, respectively. At B=18T, as seen here, the ratio reaches 50-100.

By applying high magnetic fields, one expects a magnetic field-induced semimetal-semiconductor transiton, that is, a phenomenon whereby the semimetallic superlattice state is turned into the semiconducting state. Ignoring the spin effect, Fig. 35 shows schematically the two-dimensional density of states ρ at zero magnetic field (left), at a modestly high field (middle) and at very high fields: the quantum limit (right), where ω_c is eB/m^* and $\omega_c\tau >> 1$. Similar magnetic quantization should occur for holes, too, which is not shown here. As seen in the right of Fig. 35, with increase in magnetic fields, if the ground Landau level (the peak at n=0) for electrons is raised above the Fermi level E_f, the corresponding level for holes is lowered below E_f at the same time because of the condition of the equal number of electrons and holes. This means that electrons are transferred from the InAs layer back to the GaSb layer, leaving neither conduction electrons nor holes. This transition appears to be observable with presently available dc magnetic fields for the 120-80Å superlattice whose Fermi energy is only 40meV, while the 200-100Å sample obviously requires higher magnetic fields.

Brandt and Svistova[76] claimed to have observed this type of electron transition in Bi-Sb alloys. The present superlattice system, however, has an advantage in that it provides a model much simpler and clearer for understanding the phenomenon.

iv) Far-Infrared Magneto-Absorption and Cyclotron Resonance

Far-infrared magneto-absorption experiments[77,78] were performed at 1.6K for semimetallic 120-80Å, 200-100Å and 1000-1000Å superlattices with radiation near normal incidence to the layers. Infrared sources are H_2O (λ=118μm), HCN (λ=337μm) and DCN (λ=198μm) molecular lasers, and also carcinotrons (λ=1-2μm). The transmission signal for each wavelength exhibits oscillations with increase in magnetic field. Figure 36 gives, as a function of the magnetic field B, the infrared energy positions of the transmission minima from such oscillations for the 120-80Å superlattice. The data indicate that the energies at which absorption maxima occur are directly proportional to B and all lines converge to -38$\pm$2 meV at zero magnetic field. We interpret such absorption as being due to interband transitions from H_1 to E_1 Landau levels illustrated in the inset of Fig. 36. If these transitions are assumed to occur at a selection rule, $\Delta N=0$, the converged value should correspond to the negative energy gap of the semimetallic superlattice, E_1-H_1. A similar experiment for the 200-100Å sample yielded $E_1 - H_1 = 61 \pm 4$ meV. These values are in good agreement with calculated ones.

Figure 37 shows typical transmission signals[79] as a function of magnetic field for the 1000-1000Å sample which can be considered as a series of isolated heterojunctions. Four different photon energies were used here; for each energy a number of minima, of course, correspond to absorption maxima. The most pronounced minima are attributed to cyclotron transitions of electrons, which split and become resolvable at high fields. The minima noted CR_h, which are broad and less pronounced in intensity, are believed to be associated with cyclotron transitions of holes. All other features are assigned to interband transitions. The observation of two cyclotron masses appears to manifest the multiple subband structure, as shown in the bottom of Fig. 32, where each subband has its own effective mass.

v) Polytype Superlattices

The superlattice studies in the past were limited to systems involving two host semiconductors or their pseudobinary alloys. We are now considering the introduction of a third constituent such as AlSb in the present InAs-GaSb system,[80] which provides an additional degree of freedom.

The lattice constant of AlSb (6.136Å) is compatible with those of GaSb (6.095Å) and InAs (6.058Å) for heteroepitaxy. Possible bandedge energies of AlSb relative to those of GaSb are illustrated in Fig. 38. If, indeed, the relatively wide energy-gap of AlSb (1.6eV) covers the whole range of the GaSb energy-gap and a part of the InAs energy-gap, AlSb layers serve as potential barriers for electrons as well as holes in superlattices and heterostructures. It should be realized that these three semiconductors, closely matching in lattice constant, yet significantly differing in band parameters, represent a rather unique combination among III-V compound semiconductors.

This triple-constituent system (type III) leads to a new concept of man-made polytype superlattices, ABCABC, ABAC, ACBC, etc., which can never be achieved with the dual-constituent system, as shown in Fig. 39, where A: AlSb, B: GaSb, C: InAs. Since the AlSb layers are expected to be potential barriers, the number of electrons and holes in the structure will be changed by applied electric fields. Thus, this system appears to offer an electrically-controllable medium.

Takaoka et al.[81] studied transport properties of GaSb-AlSb-InAs multiheterojunctions, one of the basic elements in the proposed superlattice. Such structures were prepared by successive MBE growths on p-type GaSb substrates with (100) surface orientation at temperatures: 500C for GaSb, 450-500C for AlSb and 450C for InAs.[82] Heteroepitaxy between GaSb and AlSb with a lattice mismatch of 0.66% appears to proceed smoothly, whereas that between InAs and AlSb with a lattice mismatch of 1.26% indicates a sign of some departure from the ideal case in the observation of streaking RHEED patterns. It is known that the small lattice mismatch is likely accommodated by relaxation involving interfacial atoms, as having been seen in the case of the InAs-GaSb superlattice,[83]

The formation of a GaSb-AlSb superlattice was confirmed metallurgically from x-ray diffraction and optically from electroreflectance and photoluminescence measurements.[84] The luminescence spectra exhibited emission peaks associated with interband transitions between quantum states as well as acceptor impurities. Voisin et al.[85] measured optical transmission near the band gap in a series of GaSb-AlSb superlattices, of which the spectra showed a staircase-like structure characteristic of the two-dimensional electron system. The absorption steps were attributed to transitions between the valence and conduction subbands in the GaSb quantum wells. The energy of the absorption edge, however, was found to be smaller than expected, which was interpreted as a result of the strain effect. Incidentally, photoluminescence from GaSb-AlSb superlattices at 300K was previously studied by Naganuma et al.[86]

vi) InAs Quantum Wells

InAs quantum wells have been investigated with MBE-grown heterostructures GaSb-InAs-GaSb and AlSb-InAs-AlSb, as shown in Fig. 40. Bastard et al.[87] performed self-consistent calculations for the electronic properties of GaSb-InAs-GaSb heterostructures including the effect of high magnetic fields, predicting the existence of a semiconductor-to-semimetal transition when the InAs thickness exceeds a threshold, as a result of electron transfer from GaSb, similar to the situation of the InAs-GaSb superlattice. This was experimentally verified by Chang et al.[88] Figures 41 and 42 show the electron densities and mobilities as a function of the InAs layer thickness, respectively, obtained from Hall measurements. In the case of GaSb-InAs-GaSb, where a single quantum well of electrons are sandwiched between two quantum wells of holes, the electron density increases as the InAs layer widens, allowing a larger number of electrons to be transferred. The mobility enhancement with increase in the thickness may be interpreted on the basis of an increase in the carrier screening effect, as well as a reduction in electron-hole scattering.

In the case of AlSb-InAs-AlSb, however, the electron density in the range of $1 \sim 2 \times 10^{12} cm^{-2}$ is unreasonably high and more or less independent on the layer thickness. Although the mobility enhancement with increase in the layer thickness is seen, it reaches a saturation value about a factor of ten lower than that in GaSb confining layers. These results lead us to assume the presence of a large number of positive charge or donor-like states in the vicinity of the interface, as illustrated in the right of Fig. 40. These states not only provide electrons but also may act as scattering centers, resulting in high electron densities and low mobilities. Neverthless, magneto-transport measurements have demonstrated the presence of two-dimensional electrons, showing pronounced oscillations at 4.2°K.

Mendez et al.[89] studied the quantized Hall effect for two-dimensional carriers confined in GaSb-InAs-GaSb heterostructures. In contrast to previously investigated systems such as GaAs-GaAlAs, the present heterostructures offer a new medium where a two-dimensional electron gas formed in InAs coexists with a two-dimensional hole gas, of the same density, formed in the GaSb side: The intrinsic nature of the bandedge relationship determined the carrier density. Figures 43 and 44 show magnetoresistance and Hall resistance as a function of magnetic field for two high-mobility samples, 252 and 262, respectively, developing well-defined Hall-resistance plateaus in good agreement with the theoretical value (h/ie^2). The magnetoresistance exhibits a typical oscillatory behavior, corresponding to the crossing of the Landau levels (N=1, 2, 3...) with the Fermi level where, at moderate fields, it is noticeable that the levels are spin-split, as indicated by the ± signs in Fig. 44. In addition to these regular characteristics, extra features are revealed, as seen at 8 and 16T for sample 252 (Fig. 43) and at 5.3, 12 and 23.5T for sample 262 (Fig. 44). An unexpected, small Hall plateau at 12T (Fig. 44), for example, was clearly distinguishable even at 4.2K, a temperature at which the fractional structures are not usually observable in the case of GaAs-GaAlAs. These extra features could result from the presence of holes and, particularly, electron-hole interaction of which the nature is not well understood at the present time.

4) Other Superlattice Structures

i) Other Semiconductor Systems

A variety of semiconductor pairs have been exploited to synthesize superlattices and heterostructures. Compared to GaAs-GaAlAs and InAs-GaSb, however, other superlattice studies are still in their infancy; in many cases, experimental results do not show as high a degree of sophistication as we have seen in the GaAs-GaAlAs system, partly because of their inferior structural quality, even though a number of intriguing proposals were made on the theoretical basis.

Nevertheless, high-quality interfaces can be obtained with InP-based lattice-matched systems. Rezek et al.[91] succeeded with a Double Heterojunction (DH) laser using InP-InGaPAs multiple thin layers. More recently, Brummell et al.[92] and Voos[93] reported magnetotransport as well as other important experiments in MOCVD-grown InP-InGaAs superlattices and heterojunctions.

Osbourn[94] proposed strained-layer superlattices from a variety of material systems including the alloys of GaAs-GaP and GaAs-InAs on a premise that lattice-mismatched heterostructures can be grown with essentially no misfit defect generation, if the layers are sufficiently thin, because the mismatch is accommodated by uniform lattice strain.[95] The GaAs-GaAsP system, which was theoretically analyzed in this proposal,[94] actually was the first superlattice experimentally investigated.[10] Ludowise et al.[95] observed stimulated emission in strained-layer GaAs-$GaAs_{0.75}P_{0.25}$ superlattices with photoexcitation. Osbourn et al.[97] (Gourley et al.[98]) claimed the growth of a direct-gap (2.03eV), strained-layer superlattice with indirect-gap hosts, GaP and $GaAs_{0.4}P_{0.6}$, due to Brillouin-zone folding. Camras et al.[99] presented data showing that Zn diffused into a strained layer GaP-$GaAs_{0.4}P_{0.6}$ superlattice enhanced the interdiffusion of As and P at the interfaces, resulting in a disordered $GaAs_{0.2}P_{0.8}$ bulk crystal. Fritz et al.[100] reported doping and transport properties of a GaAs-$Ga_{0.8}In_{0.2}As$ strained-layer superlattice. Kim et al.[101] theoretically treated GaP-AlP superlattices with varying thickness, showing that the superlattices exhibit a direct-gap behavior. Green et al.[102] and Eltoukhy et al.[103] reported the formation of GaSb-InSb periodic structures (polycrystalline) by multitarget rf sputtering, which were used for metallurgical stdudies including interdiffusion.

The Ge-GaAs system is perhaps the oldest example of heteroepitaxy[104] with nearly perfect lattice-matching. Recently, the electronic structure at this interface, which involves both polar and nonpolar faces, has drawn considerable attention for theoretical considerations: Baraff et al.,[105] Pickett et al.,[106] Harrison et al.,[107] and Pollmann et al.[108] (including Ge-ZnSe(100) interfaces). The heteroepitaxial growth of a compound semiconductor on an elemental semiconductor, however, turns out to present unsurmountable technical problems such as formation of two possible sublattices, charge neutrality, diffusion, etc. Such problems apparently prevent the achievement of the structural perfection. Nevertheless, Petroff et al.[109] and Chang et al.[110] reported the MBE growth of Ge-$Ga_{1-x}Al_xAs$ (ultra-thin layer) and Ge-GaAs superlattices, respectively, together with the results of their metallurgical analyses. Bauer et al.[111] studied surface processes controlling Ge-GaAs(100) heterojunction formation. Madhukar et al.[112] calculated the electronic structure of Si-GaP(100) interface and superlattice.

Kasper et al.[113] reported the growth of a Si-$S_{0.85}$ $Ge_{0.15}$ superlattice by UHV epitaxy with its metallurgical characterization including dislocations. Manasevit et al.[114] observed enhanced Hall mobilities in (100)-oriented, n-type Si-$Si_{1-x}Ge_x$ multilayer films.

Schulman et al.[115] theoretically treated a CdTe-HgTe superlattice, showing that its band gap is adjustable from 0 to 1.6eV depending on the thickness of the CdTe and HeTe layers. The usefulness of this superlattice for infrared optoelectric devices was suggested. Faurie et al.[116] reported the MBE growth of CdTe-HgTe superlattices which were characterized by Auger electron spectroscopy and ion microprobe profiling measurements.

Kinoshita et al.[117] successfully prepared PbTe-$Pb_{0.8}Sn_{0.2}Te$ superlattices by the hot wall technique on BaF_2 substrates, which were profiled by sputtering-Auger electron spectroscopy in conjunction with the investigation of the interdiffusion of Pb and Sn. More recently, Shubnikov-de Haas oscillations[118] were observed for two-dimensional electrons with a mobility of $1.9 \times 10^5 cm^2/Vsec$ in modulation Bi-doped superlattices. Partin[119] reported the MBE growth of PbTe-$Pb_{0.97}Ge_{0.03}Te$ superlattices with their metallurgical characterization.

ii) Ultra-Thin Layer Superlattices

Gossard et al.[120] reported the achievement of an ultra-thin layer superlattice by alternate monolayer depositions of GaAs and AlAs. Transmission electron microscopy showed such a MBE-grown structure to be perfectly epitaxial with layered composition modulation of the expected periodicity, although dark-field transmission electron micrographs indicated the presence of disordered regions along with ordered monolayer domains. The electronic properties were studied by optical absorption and luminescence. Petroff et al.[105] studied the MBE growth of Ge-$Ga_{1-x}Al_xAs$ ultra-thin layer superlattice, as previously mentioned. They also attempted alternate monolayer depositions with two components with a sizable lattice mismatch such as a pair of GaAs and InAs.

iii) Doping superlattices (n-i-p-i)

Döhler[121] and Ploog et al.[122] pursued a doping superlattice which was included in the original proposal,[1,2] as shown in the top of Fig. 3. The periodic rise and fall of the conduction and valence bands is caused by a periodic variation of impurity concentration, where electrons are attracted to minima in the conduction band while holes are attracted to maxima in the valence band. If this superlattice is illuminated, extra electrons and holes are created, which are to be spatially separated because of the above-mentioned reason. Therefore, they can have anomalously long lifetimes. An interesting consequence of this fact is that the amplitude of the periodic potential is reduced by the extra carrier, leading to a crystal which has a variable energy gap. Döhler et al.[123] indeed observed that the photon energies in luminescence, which represent the band gap, were varied from 1.3eV to 1.53eV by the laser excitation intensity. In addition, Raman experiments provided an evidence of electronic subbands in purely space-charge-induced quantum wells. The doping superlattice used here was grown by MBE, which consists of twenty n- and p-doped GaAs layers with equal doping concentration of $10^{18}cm^{-3}$. The individual layer thickness is 400Å.

IV. SUMMARY

It has been more than a decade since Esaki and Tsu, in their original proposal[2] stated, "The study of superlattices and observation of quantum mechanical effects on a new physical scale may provide a valuable area of investigation in the field of semiconductor physics." In the meantime, the evolution of MBE as a technique for the growth of ultra-thin layers of high-quality semiconductors allowed access to such a new quantum regime. Indeed, in recent years, considerable attention has been given to the engineering of artificial structures such as superlattices, quantum wells and other multiheterojunctions. Obviously, the intriguing physics, particularly regarding the electron gas system of the reduced dimensionality involved in such structures, has provided fuel for this advancement. The described InAs-GaSb research is one example which elucidates the salient features involved in synthesized superlattices. In that case, important electronic parameters such as energy gaps (including negative gaps) can be engineered by introducing made-to-order superlattice potentials: some of its unique properties may not even exist in any "natural" crystal.

We believe that efforts in this direction apparently have opened up a new area of interdisciplinary investigations in the fields of materials science and device physics. A variety of materials, including III-V, II-VI, IV-VI compounds, as well as elemental semiconductors, have been exploited for the synthesis of superlattices. In fact, the progress in the semiconductor superlattice has inspired investigations on metallic superlattices[124] and even amorphous multilayered structures.[125] In the area of semiconductor devices, ideas originating from the superlattice have found their applications in heterojunction field-effect transistors (HEMT or TEGFET), quantum-well lasers and novel avalanche photodiodes. Superlattice is now one of the important topics appearing at number of conferences and symposia, where new results are flowing in with enthusiasm. I hope this article, which cannot possibly cover every landmark, provides a little flavor to this excitement.

ACKNOWLEDGEMENTS

It is my pleasure to contribute to this special issue in the commemoration of the sixtieth birthday of my dear friend Professor Yasutada Uemura. Our investigation on superlattices was sponsored in part by the ARO.

REFERENCES

1. L. Esaki and R. Tsu, IBM Research Note RC-2418 (1969).
2. L. Esaki and R. Tsu, IBM J. Res. Develop. 14, 61 (1970).
3. D. Bohm, *Quantum Theory* (Prentice Hall, Englewood Cliffs, N.J. 1951), p. 283.
4. L. Esaki, *Les Prix Nobel en 1973*, Imprimerie Royale P. A. Norstedt & Söner , Stockholm 1974, p. 66; Science 183, 1149 (1974).
5. R. de L. Kronig and W. J. Penney, Proc. Roy. Soc. A130, 499 (1930).
6. I. I. Gol'dman and V. Krivchenokov, *Problems in Quantum Mechanics* (Addison-Wesley, Reading, Mass., 1961), p. 60.
7. R. G. Chambers, Proc. Phys. Soc. (London) A65, 458 (1952).
8. U. Gnutzmann and K. Clauseker, Appl. Phys. 3, 9 (1974).
9. A. Madhukar, J. Vac. Sci. Technol. 20, 149 (1982).
10. L. Esaki, L. L. Chang and R. Tsu, in Proceedings 12th International Conference on Low Temperature Physics, Kyoto, Japan, September 1970. (Keigaku Publishing Co., Tokyo, Japan), p.551.
11. A. E. Blakeslee and C. F. Aliotta, IBM J. Res. Develop. 14, 686 (1970).
12. A. Y. Cho, Appl. Phys. Lett. 19, 467 (1971).
13. J. M. Woodall, J. Cryst. Growth 12, 32 (1972).
14. L. L. Chang, L. Esaki, W. E. Howard and R. Ludeke, J. Vac. Sci. Technol. 10, 11 (1973); L. L. Chang, L. Esaki, W. E. Howard, R. Ludeke and G. Schul, J. Vac. Sci. Technol. 10, 655 (1973).
15. L. Esaki, L. L. Chang, W. E. Howard, and V. L. Rideout, Proceedings of the 11th International Conference on the Physics of Semiconductors, Warsaw, Poland, 1972, edited by the Polish Academy of Sciences (PWN-Polish Scientific Publishers, Warsaw, Poland, 1972), p. 431.
16. R. Ludeke, L. Esaki and L. L. Chang, Appl. Phys. Lett. 24, 417 (1974).

17. R. Tsu and L. Esaki, Appl. Phys. Lett. 22, 562 (1973).
18. L. L. Chang, L. Esaki and R. Tsu, Appl. Phys. Lett. 24, 593 (1974).
19. L. Esaki and L. L. Chang, Phys. Rev. Lett. 33, 495 (1974).
20. T. C. L. G. Sollner, W. E. Goodhue, P. E. Tannenwald, C. D. Parker, and D. D. Peck, (unpublished).
21. L. L. Chang, L. Esaki, A. Segmüller, and R. Tsu, Proceedings of the 12th International Conference on the Physics of Semiconductors, Stuttgart, Germany, 1974, B. G. Teubner, Stuttgart, p. 688.
22. A. Segmüller, P. Krishna and L. Esaki, J. Appl. Cryst. 10, 1 (1977).
23. C. Weisbuck, R. Dingle, A. C. Gossard, and W. Wiegmann, Solid State Commun. 38, 709 (1981).
24. See, for example, M. L. Cohen, Advances in Electronics and Electron Physics, Vol. 51, 1, Academic Press, 1980.
25. See, for example, Walter A. Harrison, J. Vac. Sci. Technol. 14, 1016 (1977), and also W. R. Frensley and H. Kroemer, Phys. Rev. B16, 2642 (1977).
26. G. Bastard, Phys. Rev.B, 24, 5693 (1981).
27. S. R. White and L. J. Sham, Phys. Rev. Lett. 47, 879 (1981).
28. R. Dingle, W. Wiegmann and C. H. Henry, Phys. Rev. Lett. 33, 827 (1974).
29. R. Dingle, A. C. Gossard and W. Wiegmann, Phys. Rev. Lett. 34, 1327 (1975).
30. J. P. van der Ziel, R. Dingle, R. C. Miller, W. Wiegmann, and W. A. Nordland Jr., Appl. Phys. Lett. 26, 463 (1975).
31. R. D. Dupuis and P. D. Dapkus, Appl. Phys. Lett. 32, 295 (1978).
32. N. Holonyak, Jr., R. M. Kolbas, E. A. Rezek, and R. Chin, J. Appl. Phys. 49, 5392 (1978).
33. W. T. Tsang, Appl. Phys. Lett. 39, 786 (1981).
34. R. Tsu, L. L. Chang, G. A. Sai-Halasz and L. Esaki, Phys. Rev. Lett. 34, 1509 (1975).

35. E. E. Mendez, G. Bastard, L. L. Chang, L. Esaki, H. Morkoc and R. Fischer, Phy. Rev. B26, 7101 (1982) and Physica 117B & 118B, 711 (1983).
36. P. Manuel, G. A. Sai-Halasz, L. L. Chang, C.-A. Chang and L. Esaki, Phys. Rev. Lett. 25, 1701 (1976).
37. E. Burstein, A. Pinczuk and S. Buchner, Physics of Semiconductors 1968, Institute of Physics Conference Series 43, London, 1979, p. 1231.
38. G. Abstreiter and K. Ploog, Phys. Rev. Lett. 42, 1308 (1979).
39. A. Pinczuk, H. L. Stormer, R. Dingle, J. M. Worlock, W. Wiegmann, and A. C. Gossard, Solid State Commun. 32, 1001 (1979).
40. C. Colvard, R. Merlin, M. V. Klein, and A. C. Gossard, Phys. Rev. Lett. 45, 298 (1980).
41. L. L. Chang, H. Sakaki, C. A. Chang, and L. Esaki, Phys. Rev. Lett. 38, 1489 (1977).
42. D. C. Tsui and Th. Englert, Phys. Rev. Lett. 44, 341 (1980).
43. Th. Englert and D. C. Tsui, Solid State Commun. 44, 1301 (1982).
44. H. L. Störmer, T. Haavasoja, V. Narayanamurti, A. C. Gossard, and W. Wiegmann, J. Vac. Sci. Technol. B1, 423 (1983).
45. R. Dingle, H. L. Stormer, A. C. Gossard and W. Wiegmann, Appl. Phys. Lett. 33, 665 (1978).
46. T. Mimura, S. Hiyamizu, T. Fujii and K. Nanbu, Jpn. J. Appl. Phys. 19, L225 (1980).
47. D. Delagebeaudeuf, P. Delescluse, P. Etienne, M. Laviron, J. Chaplart, and N. T. Linh, Electron. Lett. 16, 667 (1980).
48. K. v. Klitzing, G. Dorda, and M. Pepper, Phys. Rev. Lett. 45, 494 (1980).
49. D. C. Tsui and A. C. Gossard, Appl. Phys. Lett. 38, 550 (1981).
50. D. C. Tsui, A. C. Gossard, B. F. Field, M. E. Cage and R. F. Dziuba, Phys. Rev. Lett. 48, 3 (1982).

51. T. Ando and Y. Uemura, J. Phys. Soc. Jpn. 36, 959 (1974).

52. D. C. Tsui, H. L. Störmer, and A. C. Gossard, Phys. Rev. Lett. 48, 1559 (1982).

53. E. E. Mendez, M. Heiblum, L. L. Chang, and L. Esaki, to be published.

54. R. B. Laughlin, Phys. Rev. Lett. 50, 1395 (1983).

55. H. L. Störmer, A. Chang, and D. C. Tsui, Phys. Rev. Lett. 50, 1953 (1983).

56. G. Bastard, Phys. Rev. B, 24, 4714 (1981); Surf. Sci. 113, 165 (1982).

57. R. C. Miller, A. C. Gossard, W. T. Tsang, and O. Munteanu, Solid State Commun. 43, 519 (1982); R. C. Miller, A. C. Gossard, W. T. Tsang, and O. Munteanu, Phys. Rev. B 25, 3871 (1982).

58. G. Bastard, E. E. Mendez, L. L. Chang, and L. Esaki, Phys. Rev. B 26, 1974 (1982).

59. Ronald L. Greene and K. K. Bajaj, Solid State Commun. 45, 831 (1983); Ronald L. Greene and K. K. Bajaj, Solid State Commun. 45, 825 (1983).

60. H. M. Gibbs, S. S. Tarng, J. L. Jewell, D. A. Weinberger, K. Tai, A. C. Gossard, S. L. McCall, A. Passner, and W. Wiegmann, Appl. Phys. Lett. 41, 221 (1982).

61. R. Chin, N. Holonyak, Jun. G. E. Stillman, Electronics Letters, 16, 467 (1980).

62. F. Capasso, W. T. Tsang, A. L. Hutchinson, and G. F. Williams, Appl. Phys. Lett. 40, 38 (1982).

63. Federico Capasso, J. Vac. Sci. Technol. B 1, 457 (1983).

64. T. Tanoue and H. Sakaki, Appl. Phys. Lett. 41, 67 (1982).

65. Hiroyuki Sakaki, Jpn. J. Appl. Phys. 19, L735 (1980).

66. P. M. Petroff, A. C. Gossard, R. A. Logan, and W. Wiegmann, Appl. Phys. Lett., 41, 635 (1982).

67. G. W. Gobeli and F. G. Allen, in *Semiconductors and Semimetals*, edited by R. K. Willardson and A. C. Beer (Academic Press, New York, 1966), Vol. 2, p. 263.

68. H. Sakaki, L. L. Chang, R. Ludeke, C.-A. Chang, G. A. Sai-Halasz and L. Esaki, Appl. Phys. Lett. 31, 211 (1977).

69. G. A. Sai-Halasz, L. Esaki and W. A. Harrison, Phys. Rev. B18, 2812 (1978).

70. G. A. Sai-Halasz, R. Tsu and L. Esaki, Appl. Phys. Lett. 30, 651 (1977).

71. G. A. Sai-Halasz, L. L. Chang, J-M Welter, C.-A. Chang and L. Esaki, Solid State Commun. 25, 935 (1978).

72. P. Voisin, G. Bastard, C.E.T. Goncalves da Silva, M. Voos, L. L. Chang, and L. Esaki, Solid State Commun. 39, 79 (1981).

73. L. L. Chang, N. J. Kawai, G. A. Sai-Halasz, R. Ludeke and L. Esaki, Appl. Phys. Lett. 35, 939 (1979).

74. L. L. Chang, N. J. Kawai, E. E. Mendez, C.-A. Chang and L. Esaki, Appl. Phys. Lett. 38, 30 (1981).

75. N. J. Kawai, L. L. Chang, G. A. Sai-Halasz, C.-A. Chang and L. Esaki, Appl. Phys. Lett. 36, 369 (1980).

76. N. B. Brandt and E. A. Svistova, J. Low Temp Phys. 2, 1 (1970).

77. Y. Guldner, J. P. Vieren, P. Voisin, M. Voos, L. L. Chang and L. Esaki, Phys. Rev. Lett. 45, 1719 (1980).

78. J. C. Maan, Y. Guldner, J. P. Vieren, P. Voisin, M. Voos, L. L. Chang and L. Esaki, Solid State Commun. 39, 683 (1981).

79. Y. Guldner, J. P. Vieren, P. Voisin, M. Voos, J. C. Maan, L. L. Chang and L. Esaki, Solid State Commun. 41, 755 (1982).

80. L. Esaki, L. L. Chang and E. E. Mendez, Jpn. J. Appl. Phys. 20, L529 (1981).

81. H. Takaoka, C-A Chang, E. E. Mendez, L. L. Chang and L. Esaki, Physica 117B & 118B, 741 (1983).

82. C.-A. Chang, H. Takaoka, L. L. Chang and L. Esaki, Appl. Phys. Lett. 40, 983 (1982).

83. F. W. Saris, W. K. Chu, C-A Chang, R. Ludeke and L. Esaki, Appl. Phys. Lett 37, 931 (1980).

84. E. E. Mendez, C.-A. Chang, H. Takaoka, L. L. Chang, and L. Esaki, J. Vac. Sci. Technol. B1, 152 (1983).

85. P. Voisin, G. Bastard, and M. Voos, J. Vac. Sci. Technol. B1, 409, (1983).

86. M. Naganuma, Y. Suzuki, and H. Okamoto, in Proc. Int. Symp. GaAs and Related Compounds, ed. by T. Sugano (Institute of Physics, University of Reading, Berkshire, 1981) p. 125.

87. G. Bastard, E. E. Mendez, L. L. Chang, and L. Esaki, J. Vac. Sci. Technol. 21, 531 (1982).

88. C. A. Chang, E. E. Mendez, L. L. Chang, and L. Esaki, to be published.

89. E. E. Mendez, L. L. Chang, C-A. Chang, L. F. Alexander, and L. Esaki, to be published.

90. L. L. Chang, J. Vac. Sci. Technol. B1, 120 (1983).

91. E. A. Rezek, N. Holonyak, Jr., B. A. Vojak, G. E. Stillman, J. A. Rossi, D. L. Keune, and J. D. Fairing, Appl. Phys. Lett. 31, 288 (1977).

92. M. A. Brummell, R. J. Nicholas, J. C. Portal, M. Razeghi, and M. A. Poisson, Physica 117B & 118B, 753 (1983).

93. M. Voos, J. Vac. Sci. Technol. B1, 404 (1983).

94. G. C. Osbourn, J. Appl. Phys. 53, 1586 (1982).

95. J. H. van der Merwe, J. Appl. Phys., 34, 117 (1963).

96. M. J. Ludowise, W. T. Dietze, C. R. Lewis, N. Holonyak, Jr., K. Hess, M. D. Camras, and M. A. Nixon, Appl. Phys. Lett. 42, 257 (1983).

97. G. C. Osbourn, R. M. Biefeld, and P. L. Gourley, Appl. Phys. Lett. 40, 173 (1982).

98. P. L. Gourley and R. M. Biefeld, J. Vac. Sci. Technol. 21, 473 (1982).

99. M. D. Camras, N. Holonyak, Jr., K. Hess, M. J. Ludowise, W. T. Dietze, and C. R. Lewis, Appl. Phys. Lett. 42, 185 (1983).

100. I. J. Fritz, L. R. Dawson, and T. E. Zipperian, J. Vac. Sci. Technol. B1, 387 (1983).

101. J. Y. Kim and A. Madhukar, J. Vac. Sci. Technol. 21, 526 (1982).

102. J. E. Greene, C. E. Wickersham, and J. L. Zilko, J. Appl. Phys. 47, 2289 (1976).

103. A. H. Eltoukhy, J. L. Ziiko, C. E. Wickersham, and J. E. Greene, Appl. Phys. Lett. 31, 156 (1977).

104. R. L. Anderson, IBM J. Res. Develop. 4, 283 (1960); L. Esaki, W. E. Howard, and J. Heer, Appl. Phys. Lett. 4, 3 (1964).

105. G. A. Baraff, Joel A. Appelbaum, and D. R. Hamann, J. Vac. Sci. Technol. 14, 999 (1977).

106. W. E. Pickett, S. G. Louis, and M. L. Cohen, Phys. Rev. Lett. 39, 109 (1977).

107. W. A. Harrison, E. A. Kraut, J. R. Waldrop, and R. W. Grant, Phys. Rev. B18, 4402 (1978).

108. J. Pollmann and S. T. Pantelides, Phys. Rev. B21, 709 (1980).

109. P. M. Petroff, A. C. Gossard, A. Savage, and W. Wiegmann, J. Cryst. Growth 46, 172 (1979).

110. Chin-An Chang, Armin Segmüller, L. L. Chang, and L. Esaki, Appl. Phys. Lett. 38, 912 (1981).

111. Robert S. Bauer and J. C. Mikkelsen, Jr., J. Vac. Sci. Technol. 21, 491 (1982).

112. A. Madkukar and J. Delgado, Solid State Commun. 37, 199 (1981).

113. E. Kasper, H. J. Herzog and H. Kibbel, Appl. Phys. 8, 199 (1975).

114. H. M. Manasevit, I. S. Gergis, and A. B. Jones, Appl. Phys. Lett. 41, 464 (1982).

115. J. N. Schulman and T. C. McGill, Appl. Phys. Lett. 34, 883 (1979).

116. J. P. Faurie, A. Million, and J. Piaguet, Appl. Phys. Lett. 41, 713 (1982).

117. H. Konshita and H. Fujiyasu, J. Appl. Phys. 51, 5845 (1980).

118. H. Kinoshita, S. Takaoka, K. Murase, and H. Fujiyasu, Proc. of 2nd International Symposium on Molecular Beam Epitaxy and Related Clean Surface Techniques, Tokyo, Japan, p. 61.

119. D. L. Partin, J. Vac. Sci. Technol. 21, 1 (1982).

120. A. C. Gossard, P. M. Petroff, W. Weigmann, R. Dingle, and A. Savage, Appl. Phys. Lett. 29, 323 (1976).

121. G. H. Döhler, Phys. Status Solidi (b) 52, 79 and 533 (1972).

122. K. Ploog, A. Fischer, G. H. Döhler and H. Künzel, in *Gallium Arsenide and Related Compounds 1980*, Institute of Physics Conference Series No. 56, edited by H. W. Thim (Institute of Physics, London, 1981), p. 721.

123. G. H. Döhler, H. Künzel, D. Olego, K. Ploog, P. Ruden, H. J. Stolz and G. Abstreiter, Phys. Rev. Lett. 47, 864 (1981).

124. I. K. Schuller and C. M. Falco, Surf. Sci. 113, 443 (1982).

125. I. Ogino, A. Takeda and Y. Mizushima, Proceedings of the Second International Symposium on MBE and Related Clean Surface Techniques, Tokyo, Japan 1982, p. 65.

Table I SUPERLATTICE SYSTEMS

III-V/III-V	lattice mismatch	techniques	remarks
$GaAs-Ga_{1-x}Al_xAs$ (GaAs-AlAs)	0.16% for x=1	MBE,MOCVD	extensively studied
InAs-GaSb $(In_{1-x}Ga_xAs-GaSb_{1-y}As_y)$	0.61%	MBE	extensively studied
GaSb-AlSb Naganuma et al. (1981), Mendez et al. (1982), Voisin et al. (1982)	0.66%	MBE	optical properties
InAs-AlSb $(InAs-AlSb_{.84}As_{.16})$ Chang et al. (unpublished)	1.26% (~0%)	MBE	transport
$InP-In_{1-x}Ga_xP_{1-z}As_z$ x=0.12 z=0.26 Rezek et al. (1977)	~0%	LPE	D-H lasers
$InP-In_{1-x}Ga_xAs$ x=0.47 Brummell et al. (1983)	~0%	MOCVD	magneto-transport
$GaAs-GaAs_{1-x}P_x$ <0.5 Esaki et al. (1970), Ludowise et al. (1983)	1.79% for x = 0.5	MOCVD, CVD	X-ray analysis, stimulated emission
$GaP-GaP_{1-x}As_x$ <0.5 Osbourn(1982),Grouley et al.(1982),Camras et al.(1983)	1.86% for x = 0.5	MOCVD	photoluminescence, interdiffusion
$GaAs-Ga_{1-x}In_xAs$ x=0.2 Fritz et al. (1983)	1.43% for x=0.2	MBE	transport
GaP-AlP Kim et al. (1982)	0.01%	-----	theoretical
GaSb-InSb Greene et al. (1976), Eltoukhy et al. (1977)	6.29%	sputtering	interdiffusion

IV/III-V			
Ge-GaAs Chang et al. (1981)	0.08%	MBE	metallurgical
Si-GaP Mudhukar et al.(1981)	0.36%	-----	theoretical
IV/IV			
Si-$Si_{1-x}Ge_x$ $x<0.22$ Kasper et al. (1975), Manasevit et al. (1982)	0.92% for $x=0.22$	MBE, CVD	dislocations, mobility enhancement
II-VI/II-VI			
CdTe-HgTe Schulman et al.(1979), Faurie et al.(1982)	0.74%	MBE,	theoretical and metallurgical
IV-VI/IV-VI			
PbTe-$Pb_{1-x}Sn_xTe$ $x=0.2$ Kinoshita et al. (1980), (1982)	0.44% for $x=0.2$	Hot Wall	interdiffusion, magneto-transport
PbTe-$Pb_{1-x}Ge_xTe$ $x=0.03$ Partin (1982)	-----	MBE	dislocations, interdiffusion
Ultra-Thin Layer Superlattices			
GaAs-AlAs, GaAs-InAs, GaAlAs-Ge Gossard et al. (1976), Petroff et al. (1979)		MBE	metallurgical, optical properties
Doping Superlattices			
p and n GaAs Döhler (1972), Döhler et al. (1981)		MBE	extensively studied
Polytype Superlattices			
InAs-GaSb-AlSb Esaki et al. (1981)		MBE	being attempted

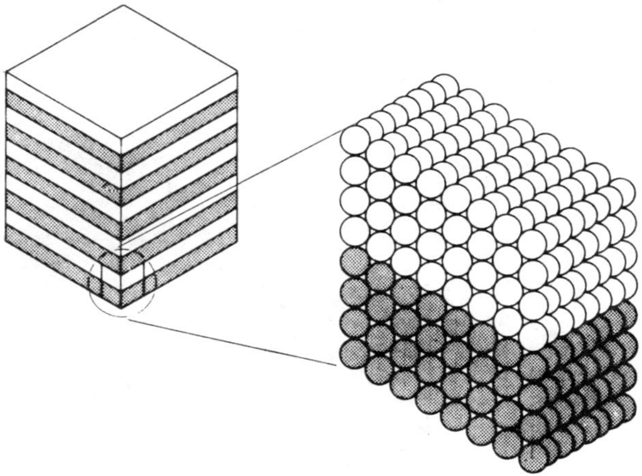

Fig. 1 Illustration of a superlattice.

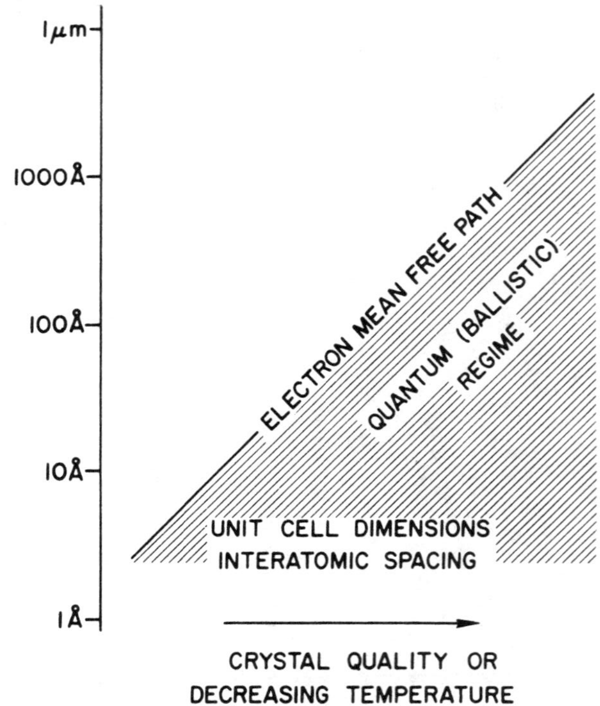

Fig. 2 Schematic illustration of a quantum regime (hatched).

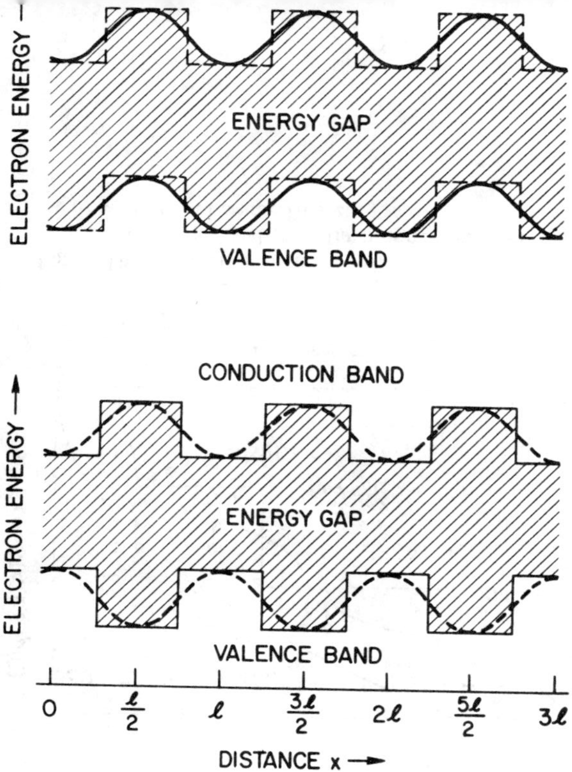

Fig. 3 Spatial variation of the conduction and valence bandedges in two types of superlattices: top, doping superlattice of alternating n-type and p-type layers; bottom, compositional superlattice with alternation of crystal composition.

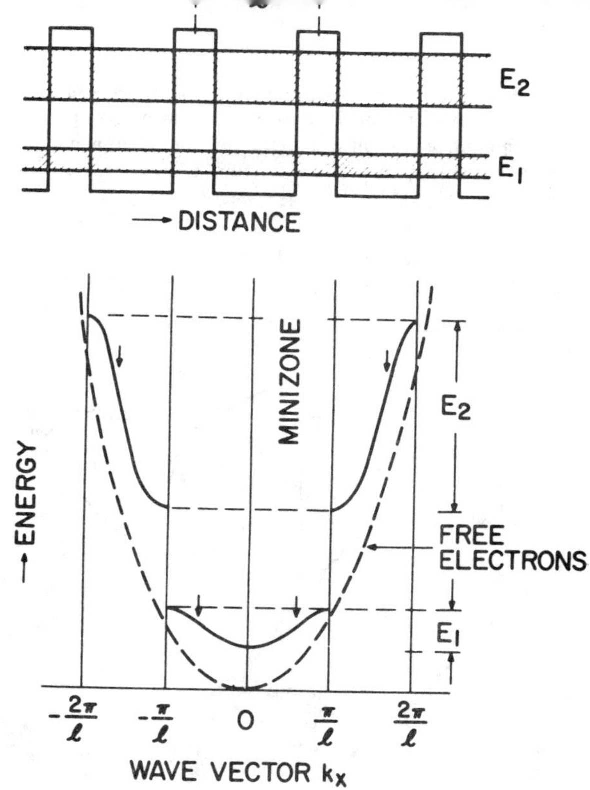

Fig. 4 Potential profile of a superlattice (top) and its energy-wave vector relationship in the minizones.

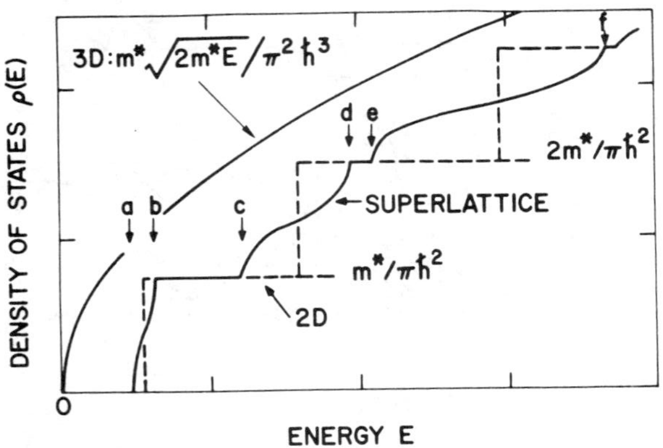

Fig. 5 The comparison of the density of states for the three-dimensional and two-dimensional electron systems with that of a superlattice.

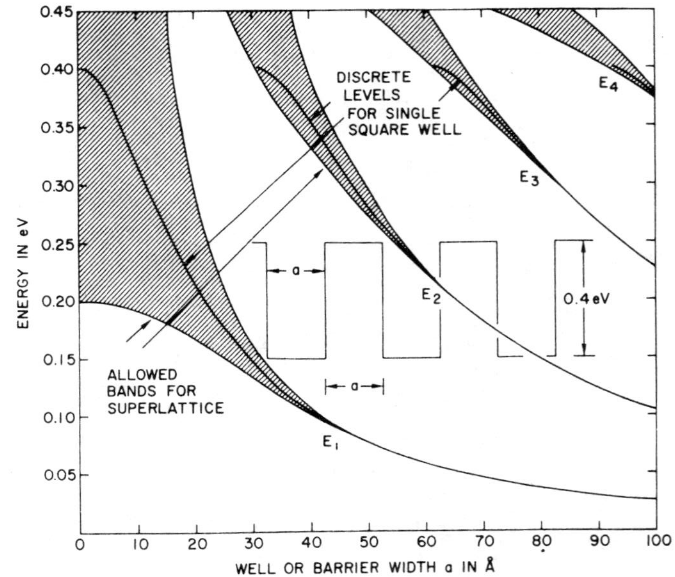

Fig. 6 Allowed energy bands, E_1, E_2, E_3 and E_4 (hatched) calculated as a function of well or barrier width, *a*, including the potential profile of a superlattice. Discrete energy levels for a single square well are also shown.

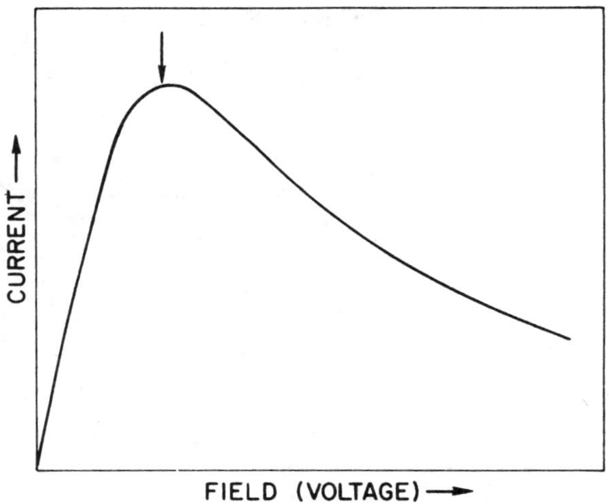

Fig. 7 A theoretical current-field (voltage) characteristic.

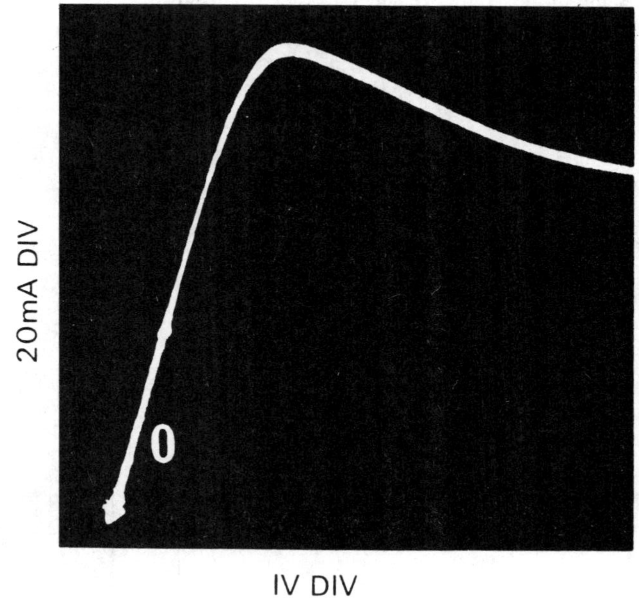

Fig. 8 Current-voltage characteristic for a 70Å period superlattice.

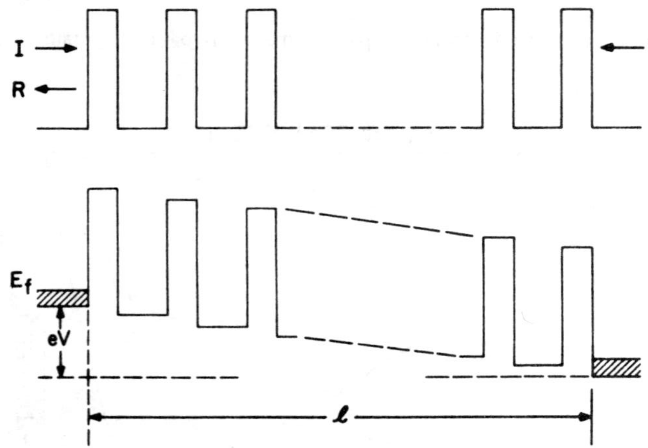

Fig. 9 Top: A finite superlattice of length ℓ, where R and T are the reflection and transmission amplitudes. Bottom: The potential profile with the application of a voltage V.

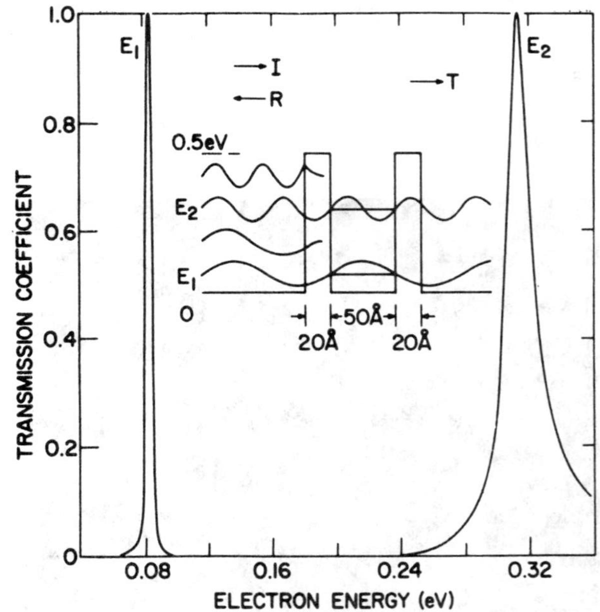

Fig. 10 Transmission coefficient T*T versus electron energy for a double barrier shown in the insert, where, if the energy of incident electrons coincides with those of the bound states, E_1 and E_2, the electrons tunnel through both barriers without attenuation.

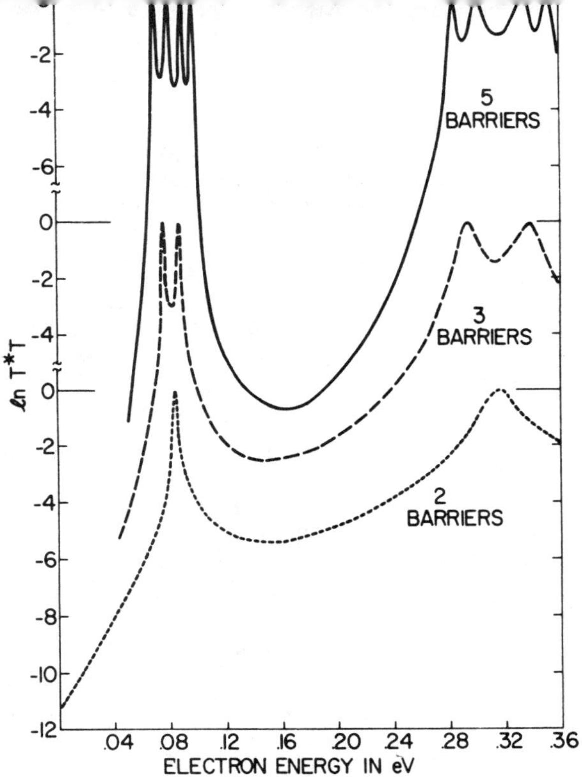

Fig. 11 Plot of *ln* T*T (transmission coefficient) versus electron energy showing peaks at the energies of the bound states in the potential wells. The curves labelled "2 barriers", "3 barriers" and "5 barriers" correspond to one, two and four wells, respectively.

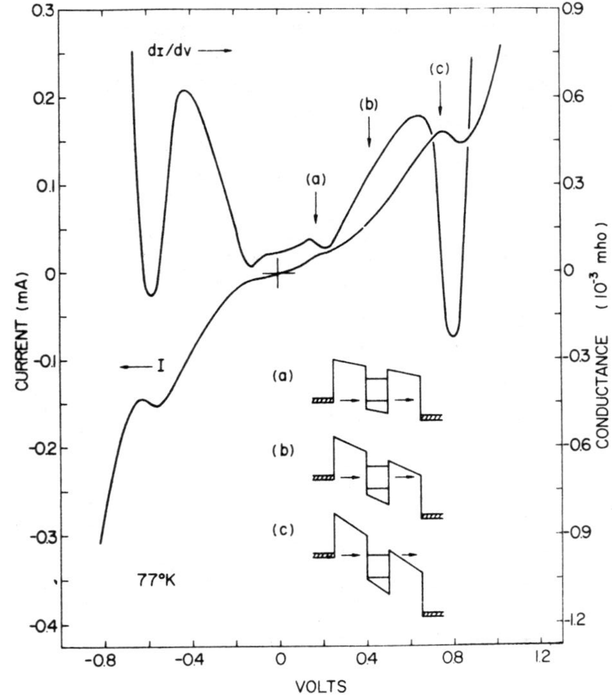

Fig. 12 Current-voltage and conductance-voltage characteristics of a double-barrier structure. Condition at resonance (a), (c) and off-resonance (b) are indicated.

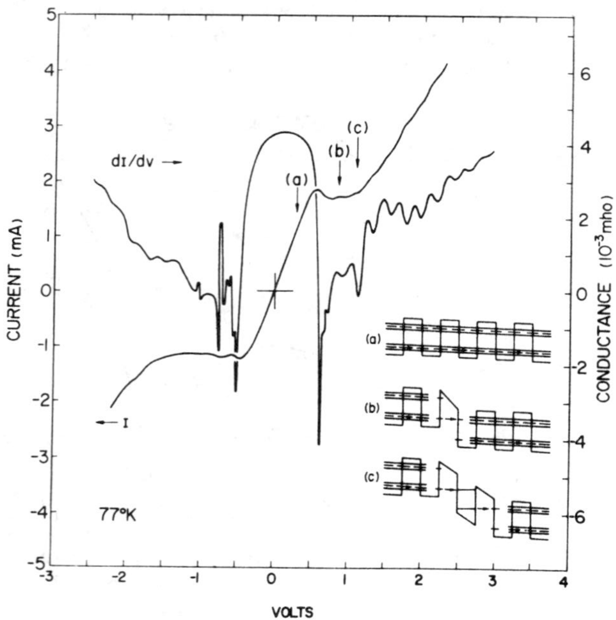

Fig. 13 Current-voltage and conductance-voltage characteristics of a superlattice. Band-type conduction (a) and localized tunneling conduction (b), (c) are indicated.

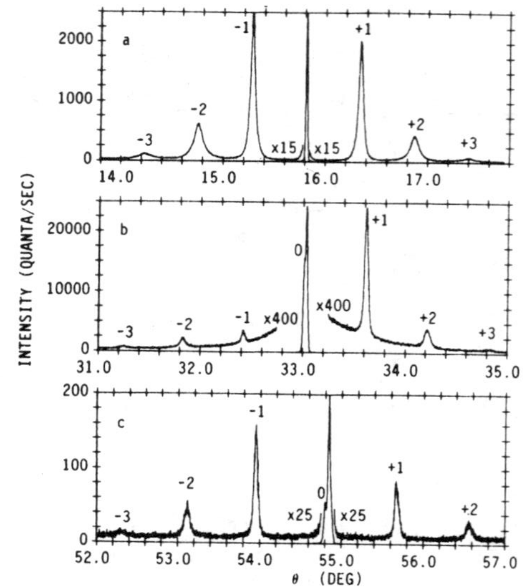

Fig. 14 High-angle diffraction patterns for a GaAs-AlAs superlattice with a period of 88Å prepared on the (001) face, obtained in the vicinity of (a) the 002, (b) the 004 and (c) the 006 reflections. Scale factors of the diffraction curves on both sides of the substrate peak are indicated.

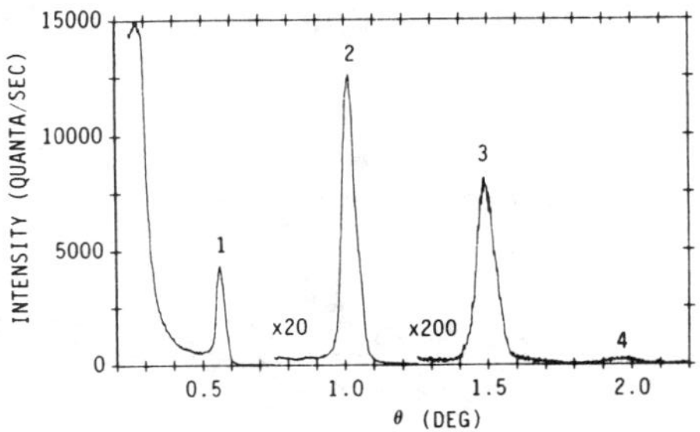

Fig. 15 Low-angle scattering pattern for a GaAs-AlAs superlattice with a period of 90.5 Å prepared on the (001) face of the GaAs substrate.

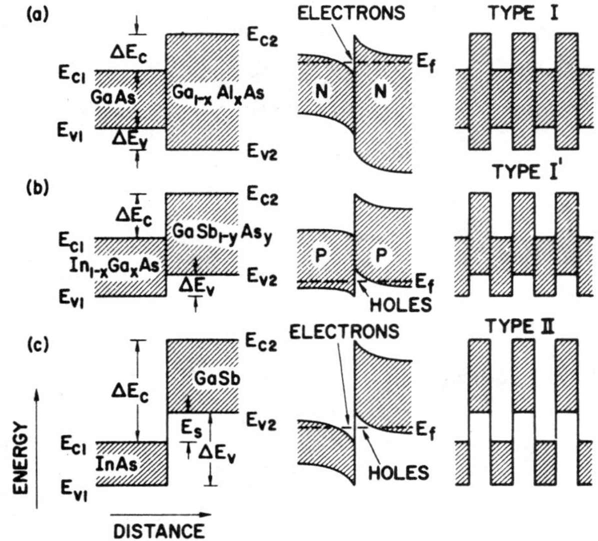

Fig. 16 The relationship of bandedge energies at the heterojunction interfaces for (a) GaAs-$Ga_{1-x}Al_xAs$, (b) $In_{1-x}Ga_xAs$-$GaSb_{1-y}As_y$ and (c) InAs-GaSb (left) and the energy band diagrams of their respective superlattices or multiquantum wells (right). The band-bending and carrier confinement at the heterojunction interfaces (middle).

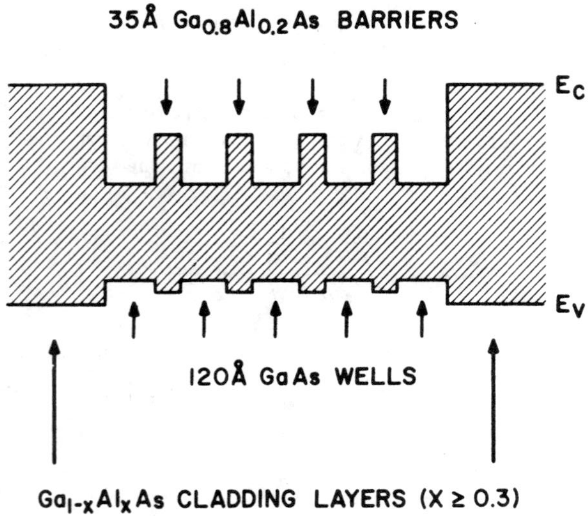

Fig. 17 Schematic energy band diagram for a multiquantum-well laser.

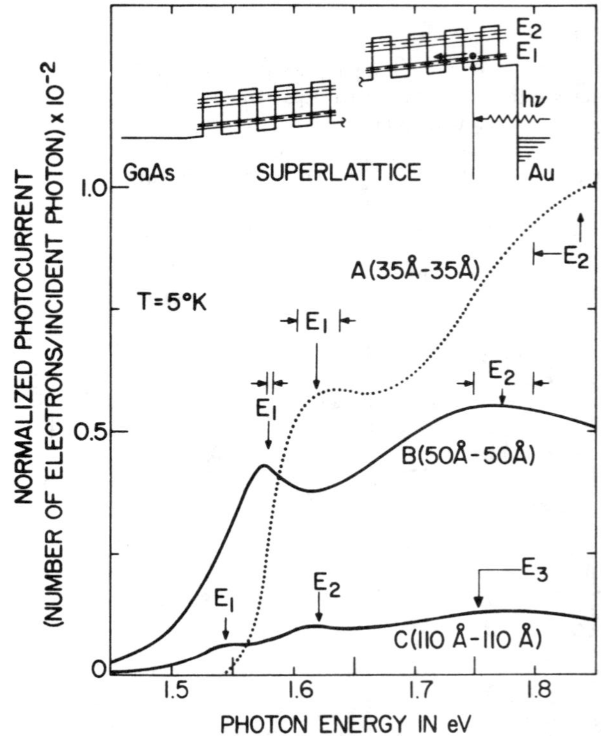

Fig. 18 Normalized photocurrent vs. photon energy for three superlattice samples A, B and C. Calculated energies and bandwidths are indicated. The energy diagram of the Au contact Schottky barrier is shown in the upper part.

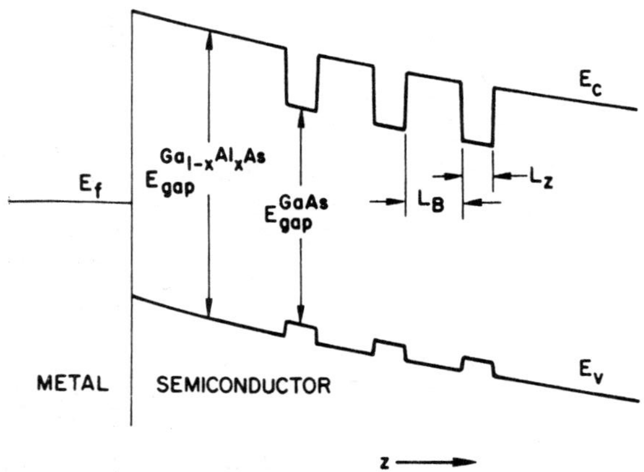

Fig. 19 Schematic energy-band diagram of a multiquantum-well Schottky junction.

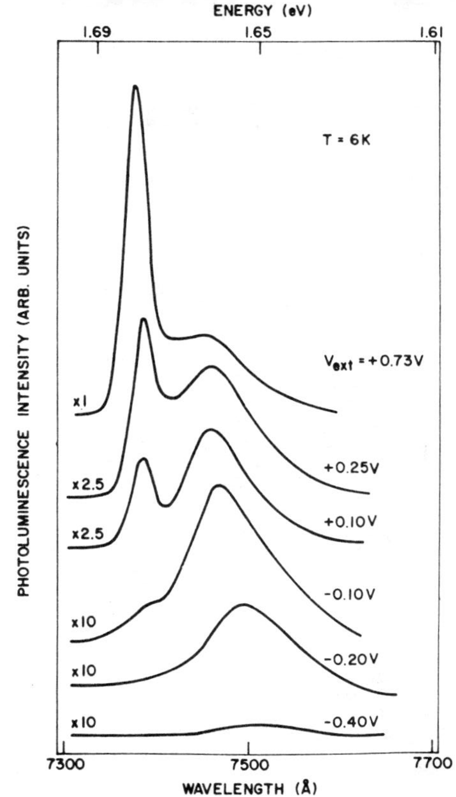

Fig. 20 Photoluminescence intensity for various voltages vs. emission wavelength for GaAs quantum wells with L_B = 35Å. The spectra were taken with 1.96eV excitation and 0.08W/cm² power density.

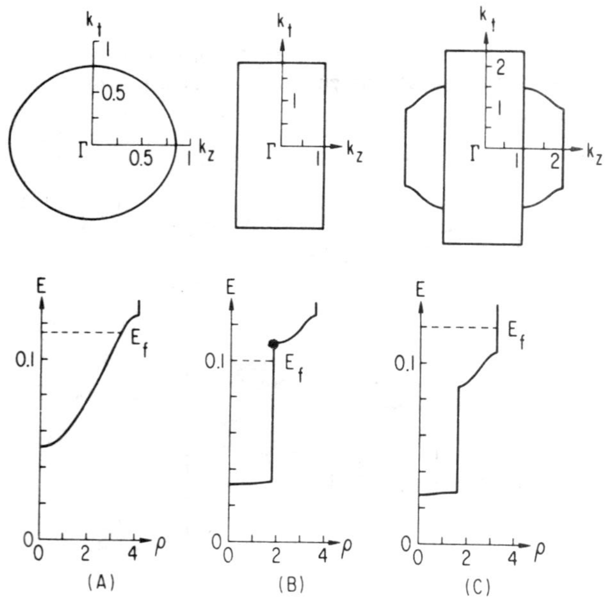

Fig. 21 Fermi surfaces (top) and the densities of states (bottom) calculated for three specimens: A, B and C.

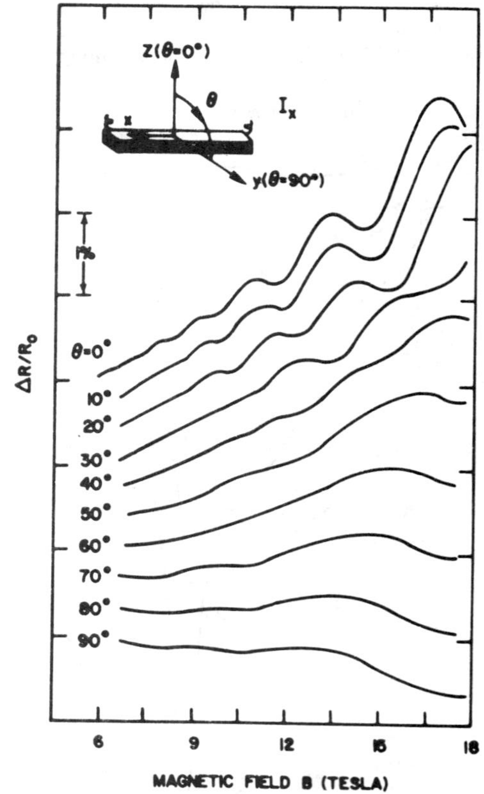

Fig. 22 Magnetoresistance versus magnetic field applied at different orientations as defined in the inset.

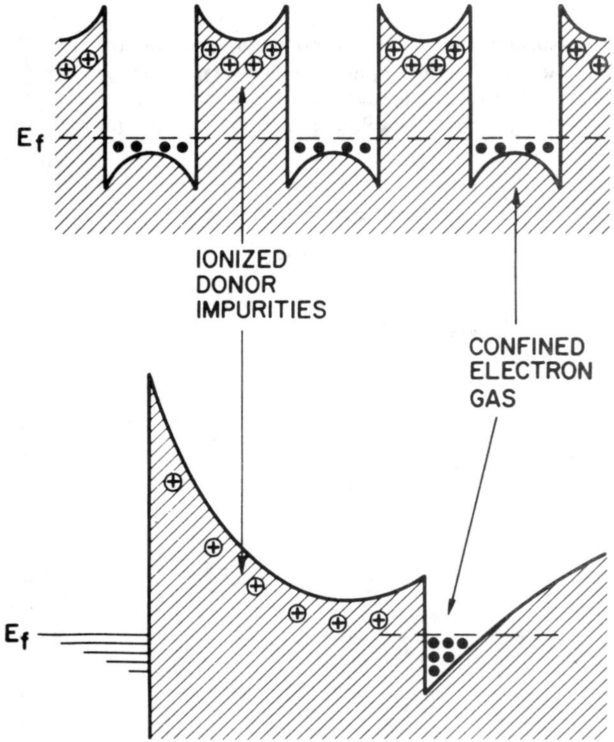

Fig. 23 Modulation doping for a superlattice (top) and a heterostructure with an attached Schottky junction (bottom).

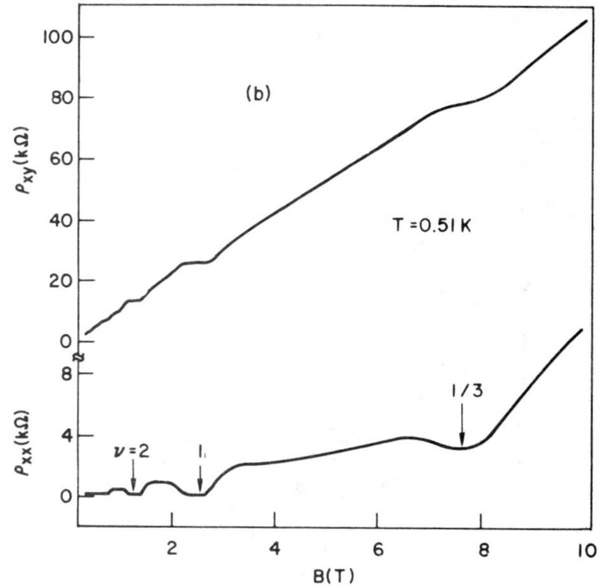

Fig. 24 Magnetoresistance ρ_{xx} and Hall resistance ρ_{xy} as a function of magnetic field at 0.51K for a heterojunction with an electron concentration of $0.6 \times 10^{11}\ cm^{-2}$

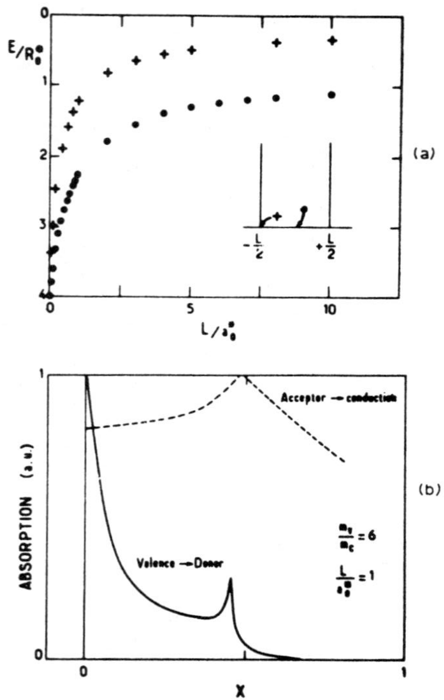

Fig. 25 (a) Binding energy E/R^*_O versus well thickness L/a^*_O, where R^*_O and a^*_O are the three-dimensional effective Rydberg and Bohr radius, repectively. (b) Absorption coefficient versus the dimensionless frequency χ.

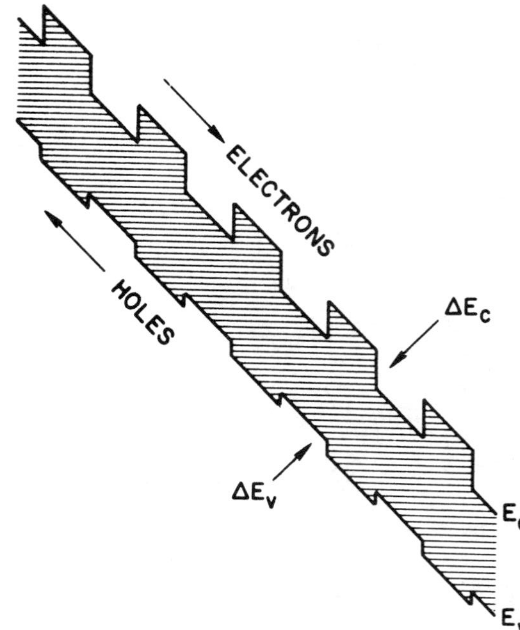

Fig. 26 Energy-band diagram of a superlattice avalanche photodiode. ΔE_c and ΔE_v are the bandedge discontinuities of the conduction and valence bands, respectively.

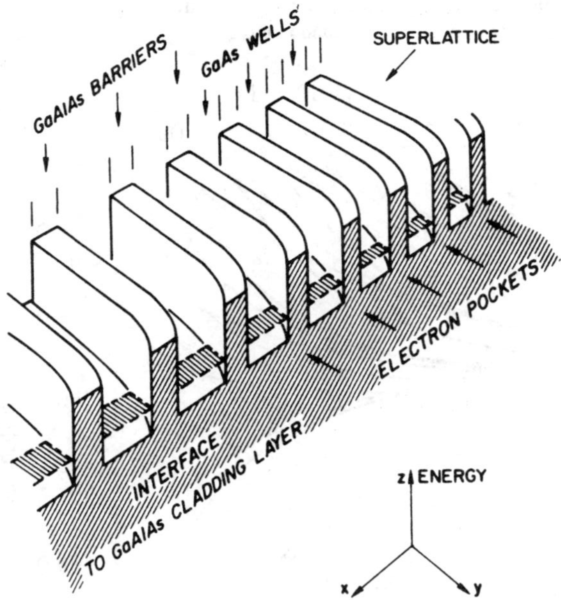

Fig. 27 The quasi-one-dimensional electron system on the surface inversion layer of a superlattice crystal.

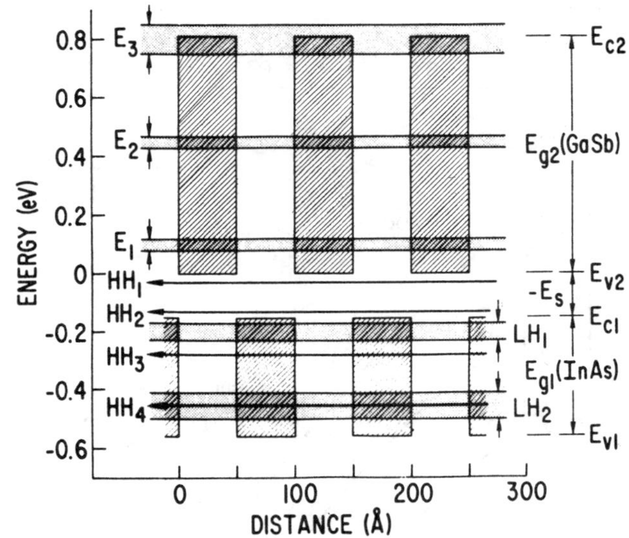

Fig. 28 Potential energy profile for an InAs-GaSb superlattice (a period of 100Å) with subband energies: E_i for electron, LH_i and HH_i for light and heavy holes, respectively.

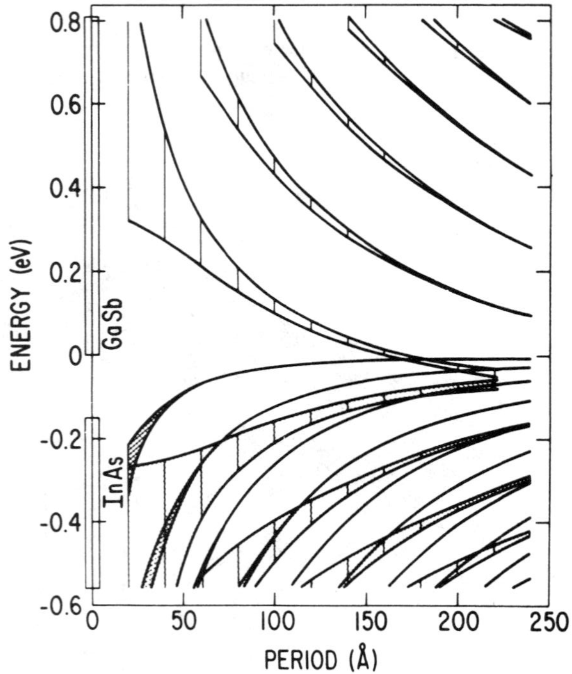

Fig. 29 Calculated subband energies and bandwidths for electrons and light and heavy holes as a function of period, assuming $d_1 = d_2$.

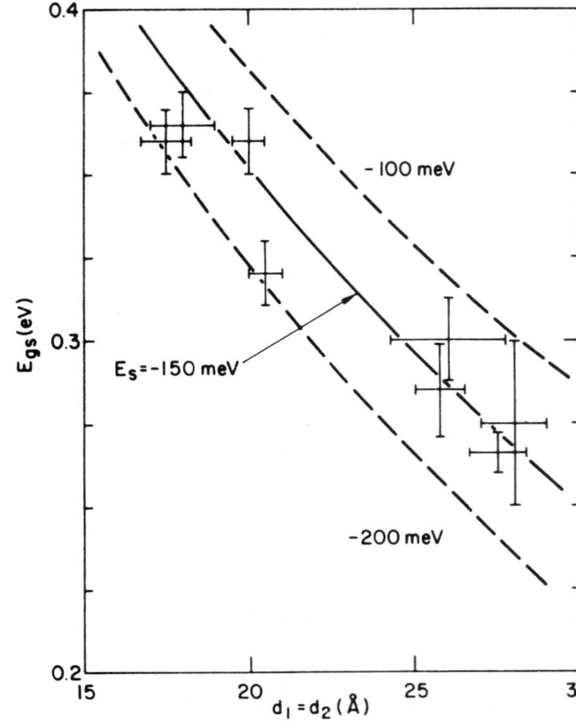

Fig. 30 Observed energy gaps as a function of the layer thickness. The energy separation E_s is used as a parameter for theoretical curves.

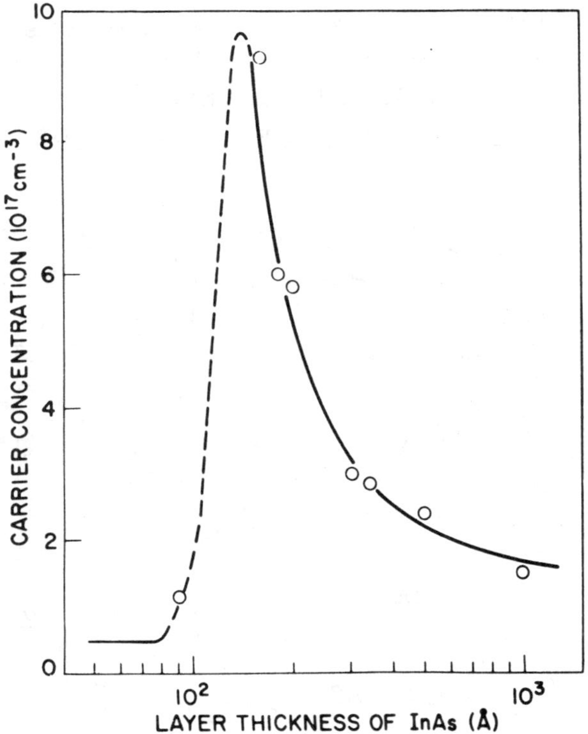

Fig. 31 Carrier concentration vs. layer thickness to demonstrate the semiconductor-semimetal transition.

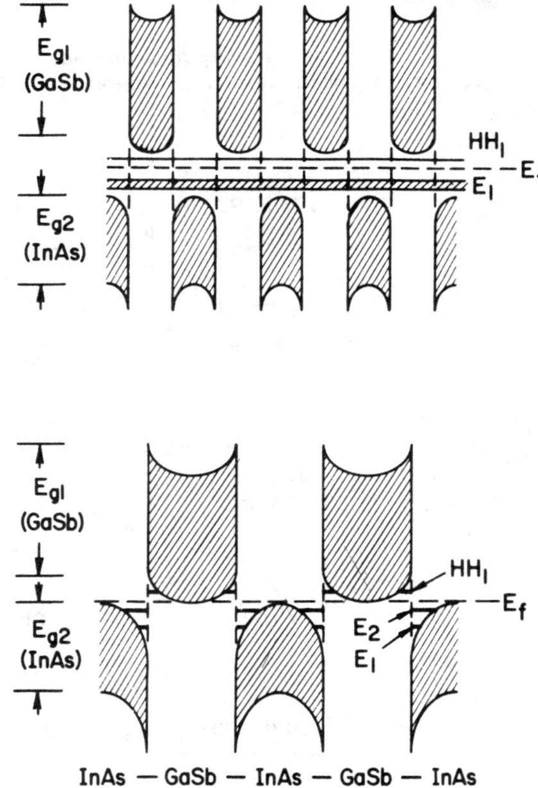

Fig. 32 Schematic illustration of potential-energy profiles for two semimetallic superlattices: thin (top) and thick (bottom) layers, where E_1 and E_2, and HH_1 are the quantized states for electrons and heavy holes, respectively.

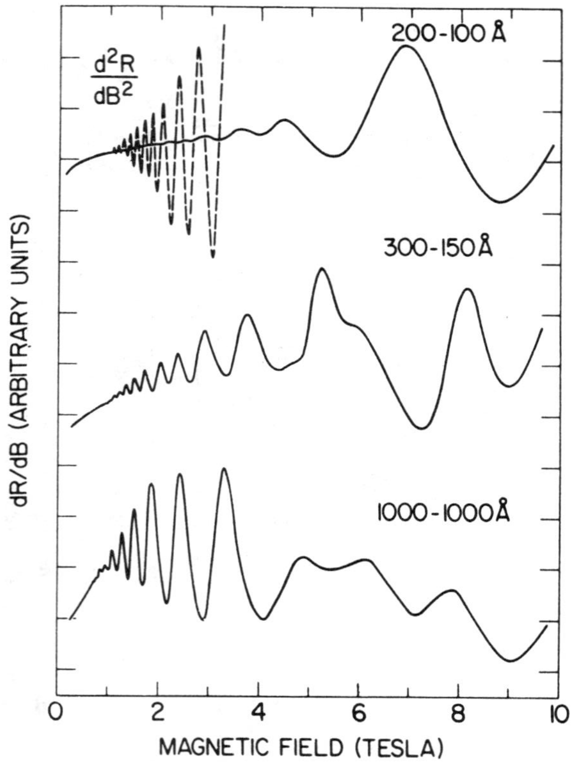

Fig. 33 Shubnikov-de Haas oscillations for three superlattices.

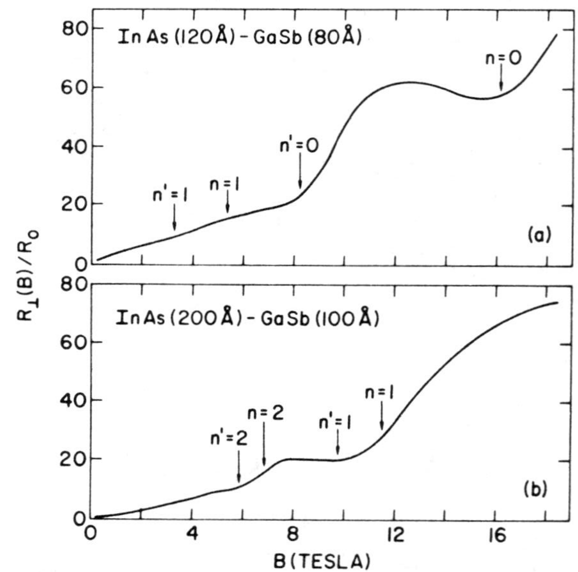

Fig. 34 Normalized transverse magnetoresistance vs. B for two superlattices.

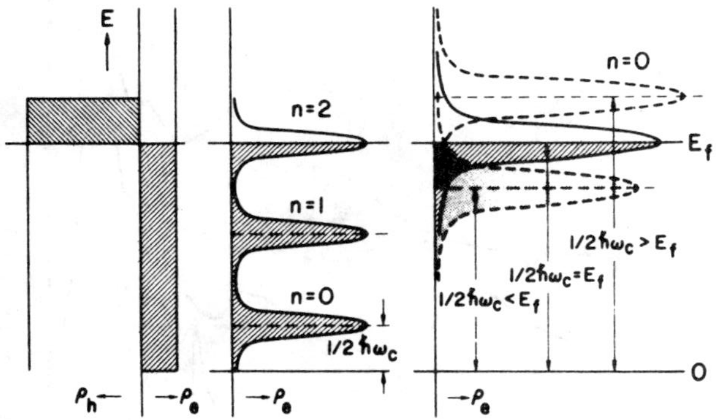

Fig. 35 Density of states at no magnetic field, at a modestly high field and at very high fields: the quantum limit.

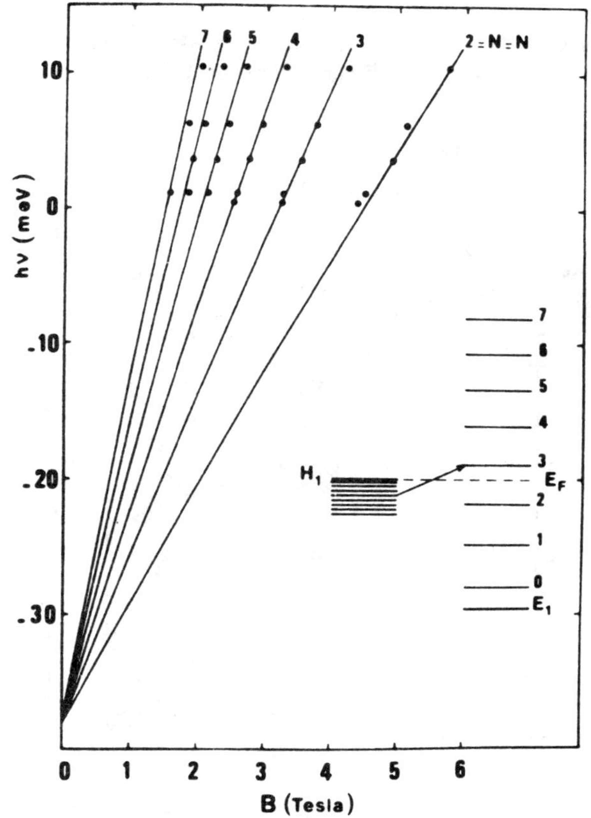

Fig. 36 Magneto-absorption energies vs. B. The inset shows the Landau levels of E_1 and H_1 to illustrate inter-band transitions.

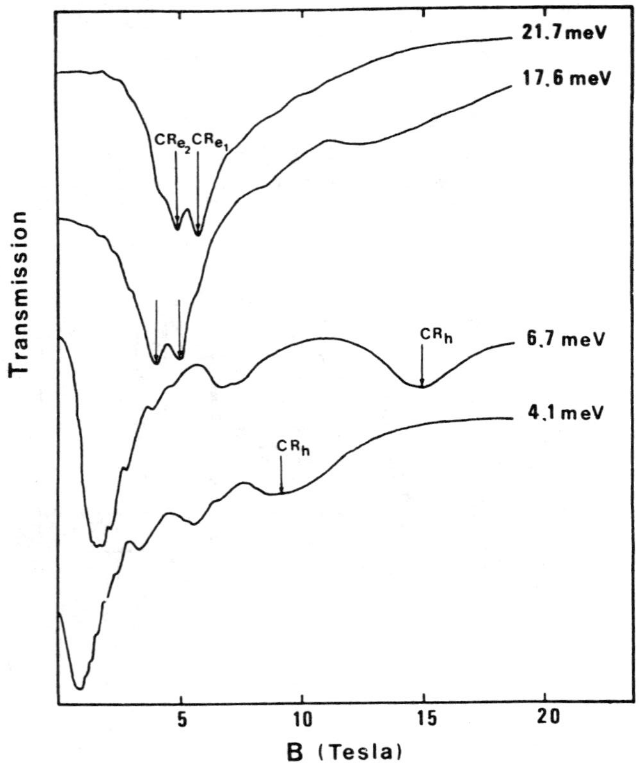

Fig. 37 Typical transmission signals vs. B for different infrared photon energies.

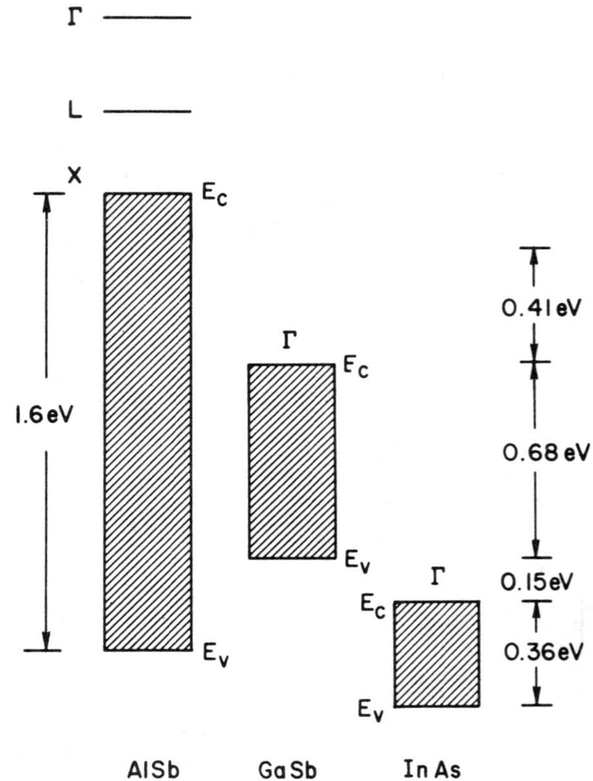

Fig. 38 Estimated bandedge energies of AlSb relative to those of GaSb and InAs. Their energy gaps are indicated by the hatched area.

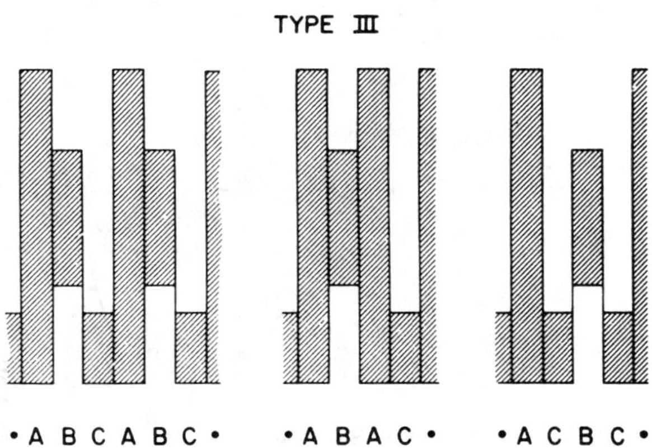

Fig. 39 Potential-energy profiles for polytype (type III) superlattices.

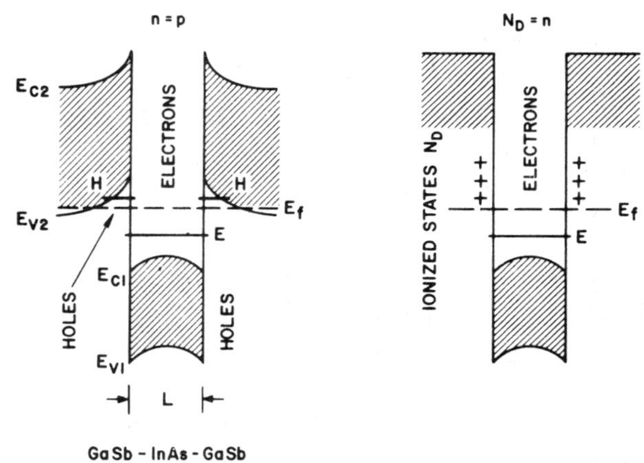

Fig. 40 Energy-band diagrams of InAs quantum wells. The left diagram is for a GaSb-InAs-GaAs heterostructure, whereas the right may be applied to an AlSb-InAs-GaAs heterostructure.

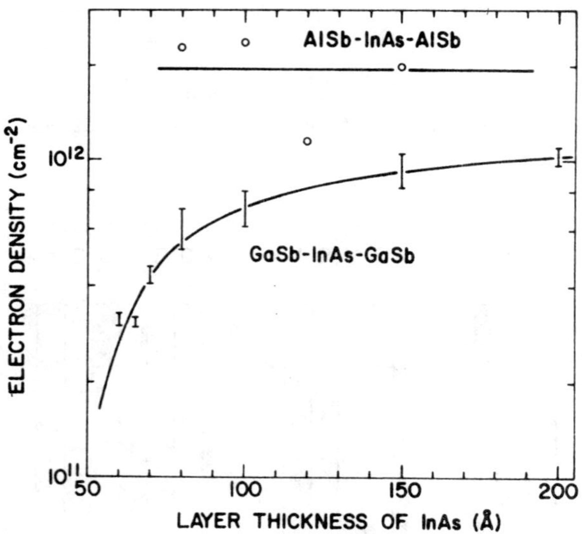

Fig. 41 Electron density at 77K versus InAs layer thickness in AlSb-InAs-AlSb and GaSb-InAs-GaSb heterostructures.

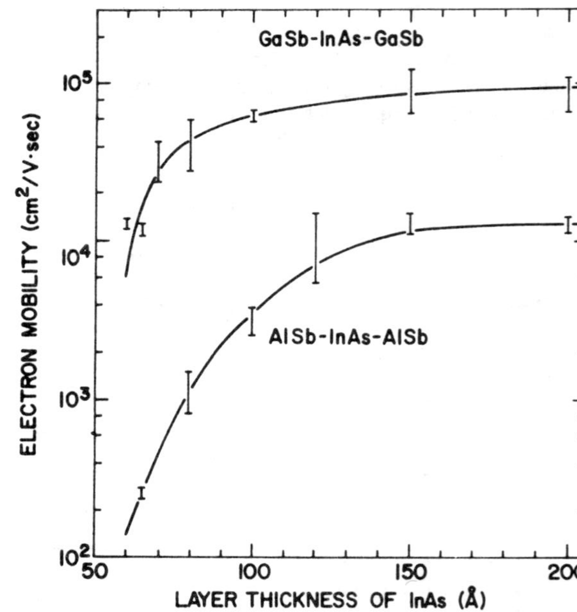

Fig. 42 Electron mobility at 77K versus InAs layer thickness in GaSb-InAs-GaSb and AlSb-InAs-AlSb heterostructures.

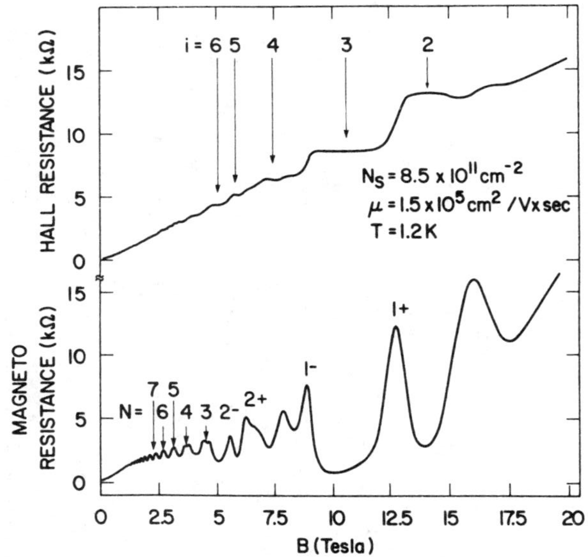

Fig. 43 Magnetoresistance and Hall resistance at 1.2K versus magnetic field for sample 252 (InAs layer thickness of 200Å) with a carrier density of $8.5 \times 10^{11}\,cm^{-2}$ and a mobility of $1.5 \times 10^5\,cm^2/Vsec$. The arrows in Hall resistance indicate a full Landau level occupation corresponding to i levels.

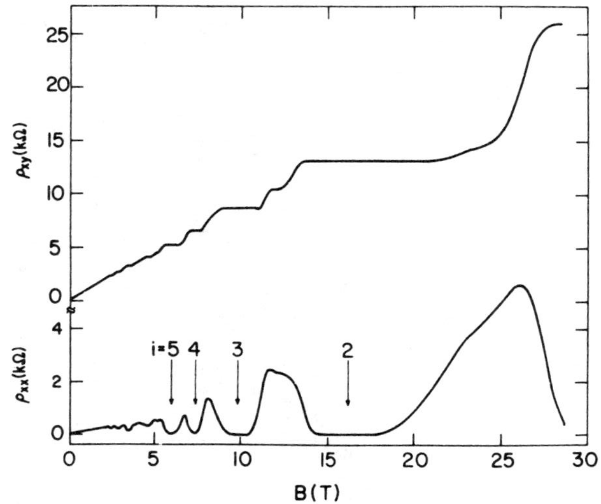

Fig. 44 Magnetoresistance ρ_{xx} and Hall resistance ρ_{xy} at 0.55K versus magnetic field for sample 262 (InAs layer thickness of 150Å) with a carrier density of $7.5 \times 10^{11}\,cm^{-2}$ and a mobility of $2.1 \times 10^5\,cm^2/Vsec$. The arrows indicate a full Landau level occupation corresponding to i levels.

QUANTIZED HALL EFFECT AND LOCALIZATION

T. Ando

Institute of Applied Physics
University of Tsukuba
Sakura, Ibaraki 305, Japan

A theoretical discussion is given on the problem of the Anderson localization in a two-dimensional system in strong magnetic fields and its relation to the quantized Hall effect. Based on results of extensive computer simulations, it is suggested that all the states except those which lie at the center of each Landau level are exponentially localized and the decay rate of localized wave functions goes to zero smoothly when the energy approaches the center of Landau levels. A brief review is given also on various investigations related to the localization in silicon inversion layers.

1. Introduction

In a two-dimensional system in strong magnetic fields the orbital motion of electrons is completely quantized and the energy spectrum comprises discrete Landau levels. The electron localization problem in the system with such a singular density of states is not only quite intriguing but also of fundamental importance. An inversion and accumulation layer made on a semiconductor-insulator interface is known to be a typical two-dimensional system. In this system, a strong external electric field and a large barrier potential of the insulator quantizes the electron motion normal to the interface, while the two-dimensional motion along the interface remaining free.[1)]

Recently, it was experimentally verified that the Hall conductivity σ_{xy} in the inversion layer made on $Si\text{-}SiO_2$ interface in strong magnetic fields (H>100 kOe) is quantized into the integer multiple of $-e^2/h$ with $h = 2\pi\hbar$ when the Fermi level lies in the localized states existing between different Landau levels.[2,3)] This is schematically illustrated in Fig. 1. A similar phenomenon has also been observed in a two-dimensional system at a $GaAs\text{-}Al_xGa_{1-x}As$ heterojunction.[4)] This fact has been suggested to provide a method of high precision measurement of

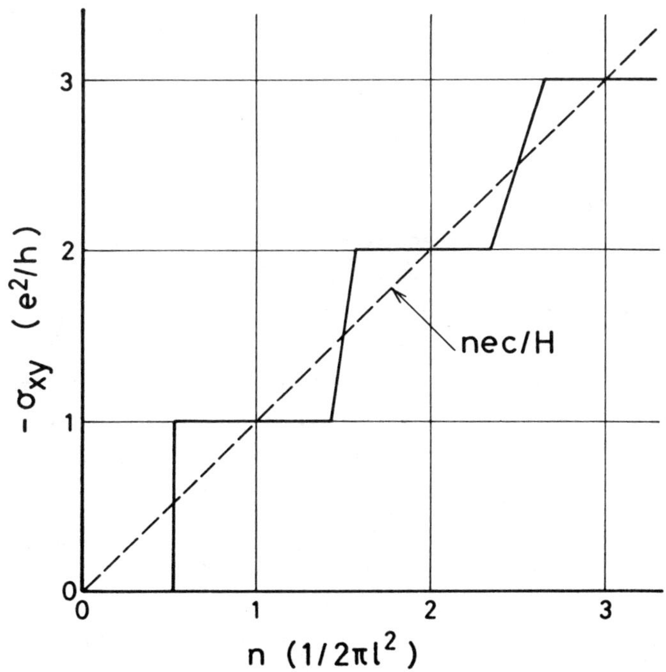

FIG. 1. The Hall conductivity is schematically shown versus electron concentration n together with a straight line representing -nec/H. The Hall conductivity is quantized into an integer multiple of $-e^2/h$ when the Fermi level lies in the energy range corresponding to localized states.

the universal quantitity e^2/h,[2)] and actually been used to determine e^2/h within an error of 10^{-7}.[5)] This so-called quantized Hall effect has a close relationship with the electron localization and/or delocalization. The purpose of this paper is to give from a theoretical point of view a review on the quantized Hall effect and the electron localization in the two-dimensional system in strong magnetic fields.

A discussion on some exact properties of the Hall conductivity will be given in Sec. 2 with emphasis on their relation to the electron localization. It will be shown that in strong magnetic fields all the states cannot be localized within the Landau level. In Sec. 3 various investigations related to the electron localization in silicon inversion layers are reviewed briefly. In Sec. 4 results of extensive numerical experiments on the Anderson localization are presented. The numerical results suggest that all the states except at the center of each Landau level are localized exponentially. Section 5 is used for discussions of some of new or unresolved problems.

2. Quantized Hall Effect

2.1 Historical Survey

When an electric field E_y is applied in the y direction, each electron moves in the x direction with a constant velocity $v_x = cE_y/H$ inversely proportional to the magnetic field H. This gives rise to the well-known expression of the Hall conductivity $\sigma_{xy} = -nec/H$, where n is the concentration of electrons in a unit area, and also the vanishing diagonal conductivity σ_{xx}. In the presence of scatterers like impurities, a frictional force acting on each electron modifies the drift motion of electrons and the Hall conductivity is no longer given by this expression. Further, the diagonal conductivity becomes nonzero.

In case of short-range scatterers the peak value of σ_{xx} has been calculated in the self-consistent Born approximation and its peak value is given by $(N+1/2)\ e^2/\pi^2\hbar$.[6)] This has been confirmed by various experiments in the silicon inversion layer.[7,8)] Ando, Matsumoto, and Uemura calculated the Hall conductivity in several approximations.[9)] In the self-consistent Born approximation, σ_{xy} at T=0 has been shown to be related to σ_{xx} by

$$\sigma_{xy} = -\frac{nec}{H} - \frac{\mathrm{Im}\Sigma(E_F)}{\hbar\omega_c}\sigma_{xx} \quad , \tag{2.1}$$

where $\Sigma(E)$ is the self-energy due to impurity scatterings, E_F is the Fermi energy, and ω_c (=eH/mc) is the cyclotron frequency. This relation can be understood as follows. When an electric field E_y is applied in the y direction, the center of the cyclotron orbit moves in the x direction with a drift velocity $v_x = cE_y/H$, which gives rise to the Hall conductivity -nec/H. Effects of scattering can be regarded as a frictional force acting on each electron, whose strength is given by $F_x = -mv_x/\tau_f = eE_y/\tau_f$, where τ_f is a relaxation time, i.e., $\tau_f \sim -\hbar/2\mathrm{Im}\Sigma(E)$ in case of short-range scatterers. The current in the x direction due to this force is given by

$$\Delta j_x = \sigma_{xx}\frac{F_x}{(-e)} = \frac{1}{\omega_c\tau_f}\sigma_{xx}E_y \quad . \tag{2.2}$$

One of the important conclusion reached by Ando et al. is that the Hall conductivity is given by $\sigma_{xy} = -Ne^2/h$ when the Fermi level lies in the energy gap between the (N-1)th and Nth Landau levels. Since the total density of states of each Landau level is $1/2\pi\ell^2$, where $\ell^2 = c\hbar/eH$, this leads to the interesting conclusion that the Hall conductivity when each Landau level is completely filled is not affected by the presence of impurities. This fact was confirmed experimentally by Kawaji, Igarashi, and Wakabayashi.[10] However, this interesting phenomenon has not attracted much interest until quite recently, partly because of experimental difficulties in accurately measuring the off-diagonal conductivity.

Prange[11] recently considered the problem of a single impurity with a short-range delta potential and reached the same conclusion obtained by Ando et al. Aoki and Ando[12] elucidated some exact properties of σ_{xy} in strong-field limits where each Landau level can be treated separately. A theoretical argument which leads to the quantized Hall effect for an arbitrary strength of the magnetic field has been provided by Laughlin.[13] His theory has subsequently been supplemented by Halperin.[14] According to Laughlin's argument the Hall conductivity should be an integer (can be zero) times $-e^2/h$ whenever the Fermi level lies in localized states and there is a nonzero gap in the energy spectrum for extended states. Some properties of the Hall conductivity were reported also by Thouless.[15] In the following, the theory of Aoki and Ando is reviewed first with emphasis on the relation between the quantized Hall effect and the problem of the electron localization. Then, a different version of Laughlin's argument is presented to explain the value of the quantized Hall conductivity in arbitrary magnetic fields.

2.2 Effect of Localization on the Hall Conductivity

The Hamiltonian of a two-dimensional system in a magnetic field of strength H with random potential V(r) is given by

$$H = \frac{1}{2m}\pi^2 + V(r) = \frac{1}{2m}(p+\frac{e}{c}A)^2 + V(r) \quad , \tag{2.3}$$

where A is a vector potential. The system is assumed to be in an xy plane with the linear dimension L. The magnetic field is in the z direction perpendicular to the system. A periodic boundary condition is imposed in the y direction. The Landau wave function $|NX)$ being used as a basis, the above can be written as

$$H = \sum_{NX} E_N |NX)(NX| + \sum_{NX} \sum_{N'X'} |NX)(NX|V|N'X')(N'X'| \quad , \tag{2.4}$$

where N is the Landau-level index starting with 0, X the eigenvalue of a center coordinate X of the cyclotron motion, and $E_N = (N+1/2)\,\hbar\omega_c$. The basis wave function is given in the gauge $A = (0,Hx)$ by

$$\phi_{NX}(r) = (2^N N!\sqrt{\pi}\ell)^{-1/2}\, H_N(\frac{x-X}{\ell})\exp[-\frac{1}{2}(\frac{x-X}{\ell})^2]\,\frac{1}{\sqrt{L}}\,\exp(-i\frac{Xy}{\ell^2}) \quad , \tag{2.5}$$

where $H_N(t)$ is Hermite's polynomial and X takes the discrete value $j\Delta X$ with $j = 1,2,..,N_X$ owing to the periodic boundary condition in the y direction. We have defined $\Delta X = 2\pi\ell^2/L$ and $N_X = L^2/2\pi\ell^2$. As has been shown by Kubo, Hashitsume, and Miyake,[16] the coordinate (x,y) of an electron can be expressed as a sum of relative (ξ,η) and center coordinates (X,Y) by $x = \xi + X$ and $y = \eta + Y$ with $\xi = (\ell^2/\hbar)\,\pi_y$ and $\eta = -(\ell^2/\hbar)\,\pi_x$. In the absence of $V(r)$, the Hamiltonian contains only the relative coordinates and each Landau level has the degeneracy corresponding to the position of the center of the cyclotron orbit. The comutation relation $[X,Y] = i\ell^2$ gives the well-known degeneracy $L^2/2\pi\ell^2$ of the Landau level. In the presence of the impurity potential $V(r)$, the center coordinates are no longer a constant of motion, and the density of states of each Landau level have a nonzero width. This problem has been a subject of an extensive theoretical study. See, for example, Sec. VI of ref. 1 for a detailed discussion.

The Hall conductivity is expressed as[16]

$$\sigma_{xy} = -\frac{nec}{H} + \Delta\sigma_{xy} \quad , \tag{2.6}$$

where n is the electron concentration and $\Delta\sigma_{xy}$ is given by the correlation function of the velocities of the center coordinates X and Y as

$$\Delta\sigma_{xy} = \frac{e^2\hbar}{iL^2} \langle \sum_\alpha f(E_\alpha) \sum_\beta \frac{P}{(E_\alpha - E_\beta)^2}$$

$$\times [(\alpha|\dot{X}|\beta)(\beta|\dot{Y}|\alpha) - (\alpha|\dot{Y}|\beta)(\beta|\dot{X}|\alpha)] \rangle \quad , \qquad (2.7)$$

with f(E) being the Fermi distribution function, and $|\alpha)$ the eigenstate of the Hamiltonian (2.3) with eigenenergy E_α. The velocities are given by

$$\dot{X} = \frac{\ell^2}{\hbar} \frac{\partial V(r)}{\partial y} \quad , \qquad \dot{Y} = -\frac{\ell^2}{\hbar} \frac{\partial V(r)}{\partial x} \quad , \qquad (2.8)$$

The expression (2.6) with (2.7) immediately leads to $\sigma_{xy} = -nec/H$ in the absence of V(r). The second $\Delta\sigma_{xy}$ describes corrections due to scatterings by impurities and also localization effects.

If $|\alpha)$ is a localized state, we have, for any $|\beta)$,

$$(\alpha|\dot{X}|\beta) = \frac{1}{i\hbar}(E_\beta - E_\alpha)(\alpha|X|\beta) \quad . \qquad (2.9)$$

Therefore, the contribution from the state $|\alpha)$ to $\Delta\sigma_{xy}$ in (2.7) reduces to $\Delta\sigma_{xy}{}^\alpha = f(E_\alpha)ec/H$, where use has been made of $[X,Y] = i\ell^2$. From this simple result we can derive important general properties at T = 0:

(i) As long as the Fermi level lies in the localized regime of the energy spectrum, σ_{xy} remains constant independent of n.

(ii) If all the states below the Fermi level E_F are localized, we have $\sigma_{xy} = 0$ because $\Delta\sigma_{xy}$ exactly cancels -nec/H in (2.6).

Let us now study the limit of strong magnetic fields, where the energy spectrum comprizes separate Landau levels and mixings between different Landau levels can be completely neglected. In this case there is a situation in which Landau levels are just filled up to the Nth level when E_F lies in the gap between the Nth and (N+1)th levels with $n = N/2\pi\ell^2$. Now it can be shown that

(iii) $\Delta\sigma_{xy} = 0$, i.e., $\sigma_{xy} = -Ne^2/h$ in this case. This is readily seen from (2.7) if we note that $f(E_\alpha)$ can be replaced by $f(E_\alpha)[1-f(E_\beta)]$ in that equation. It follows that, in the limit of strong magnetic fields,

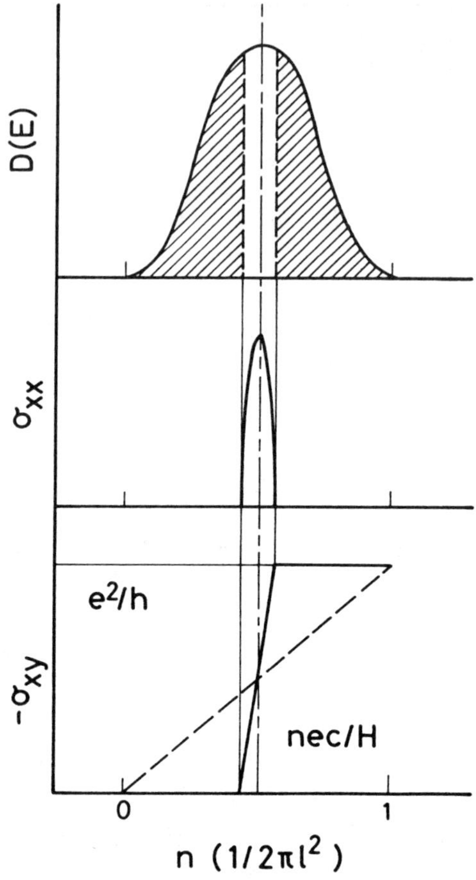

FIG. 2. The density of states D(E), the diagonal conductivity σ_{xx}, and the Hall conductivity σ_{xy} are schematically shown as a function of the electron concentration n for a single Landau level. Shaded parts in the density of states correspond to localized states.

(iv) σ_{xy} assumes the constant value $-Ne^2/h$ when E_F lies in the localized regime between the Nth and (N+1)th levels according to the properties (i) and (ii). Combining the properties (ii) and (iii), we can also draw an important conclusion that

(v) all the states cannot be localized in the two-dimensional system in strong magnetic fields.

Further, using a symmetry property of the Hamiltonian we can show that

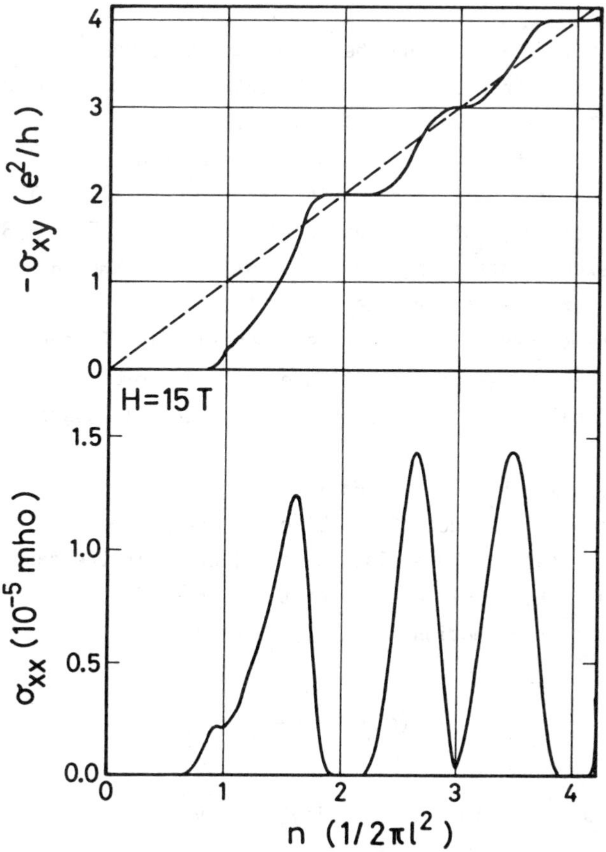

FIG. 3. An example of the diagonal conductivity (lower panel) and the Hall conductivity (upper panel) observed in a Si inversion layer by Kawaji and Wakabayashi.[3] The dashed line represents nec/H. Only the lowest N=0 Landau levels (split into four by spin-Zeeman and valley splittings) are shown.

$$\Delta\sigma_{xy}(V;H;\nu) = \Delta\sigma_{xy}(-V;-H;1-\nu) = -\Delta\sigma_{xy}(-V;H;1-\nu) \ , \qquad (2.10)$$

where ν is the occupation ratio of the Landau level defined by $\nu = 2\pi\ell^2 n_N$ with n_N being the electron concentration of the level. By this formula we can relate σ_{xy} in the lower edge of a Landau level to that in the upper edge of the level. Especially in case that the system contains equal amounts of attractive and repulsive scatterers, (2.10) gives a symmetry of σ_{xy} around the center of the Landau level. So far, effects of electron-electron interactions have been completely neglected. The last symmetry relation (2.10) can be shown to be valid in the presence of interactions among electrons, leading to the quantization of the Hall conductivity in this case also. The obtained results are schematically illustrated in Fig. 2. Figure 3 gives an example of experiments observed by Kawaji and Wakabayashi,[3] who demonstrated experimentally the correctness of the characteristic feature of the theoretical expectation shown in Fig. 2 for the first time. The figure seems to indicate that the lowest Landau level is completely localized. This problem will be discussed further in Sec. 5.3.

It has been shown that the Hall conductivity assumes the value $-Ne^2/h$ despite the presence of immobile carriers. The physical meaning of this property can be explained as follows. In the case of strong magnetic field, the matrix element of the current j_x reduces to that of $-e\dot{X}$. The equation of motion for X is given as

$$\dot{X} = \frac{\ell^2}{\hbar}\,(eE_y + \frac{\partial V(r)}{\partial y}) \quad . \tag{2.11}$$

Since σ_{xy} is a nondissipative quantity, we have

$$j_x = -\frac{e}{H}\sum_{\alpha'}[eE_y + (\alpha'|\frac{\partial V}{\partial y}|\alpha')] \quad , \tag{2.12}$$

where the summation for α' is over the occupied states and we have

$$\phi_{\alpha'}(r) = \sum_X c_{NX}^{\alpha'}\,\phi_{NX}(r) \quad . \tag{2.13}$$

Note that the wave function $|\alpha')$ contains effects of the infinitesimal external electric field E_y. If all the states associated with the Landau level are occupied, we have

$$\sum_{\alpha'} (\alpha'|\frac{\partial V}{\partial y}|\alpha') = \int dr \sum_X |\phi_{NX}(r)|^2 \frac{\partial V(r)}{\partial y} = \frac{1}{2\pi\ell^2} \int dr \frac{\partial V(r)}{\partial y} = 0 \quad , \qquad (2.14)$$

where use has been made of

$$\sum_X |\phi_{NX}(r)|^2 = \frac{1}{2\pi\ell^2} \; . \qquad (2.15)$$

This clearly leads to the conclusion (iii) mentioned above.

For a localized state, the effective field of the random potential modified by a slight deformation of the wave function cancels the applied electric field, and the localized state does not contribute to the Hall conductivity. On the other hand, electrons in extended states feel an effective field in the direction of the external field and move faster because of the severe constraint (2.15) and because their wave functions should be orthogonal to those of localized states. The fast motion of electrons in the extended states exactly cancels the decrease of the current due to localized electrons.

The above mentioned reasoning of the fact that the Hall conductivity is $-e^2/h$ despite the presence of localized electrons becomes obvious when the range of fluctuating potentials is much larger than the radius of the cyclotron orbit. In this case the electron motion is almost classical and the center of the cyclotron orbit moves along an equipotential line with a velocity proportional to the spatial derivative of the local potential energy. As is clear in Fig. 4, the total current in the direction perpendicular to the applied electric field is obtained by integrating the potential derivative from the lower edge to the upper edge of the system. The result is always exactly equal to the potential drop between the system edges independent of whether fluctuating potentials are present or not.

We have a nonzero σ_{xy} in the strong-field limit, which shows that all the states within a single Landau level are not localized in this system. This indicates that the conclusion reached by Abrahams, Anderson, Licciardello, and Ramakrishnan[17)] (the so-called gang of four) that all the states are localized in the two-dimensional system is not applicable to the present system in strong magnetic fields. This important conclusion in strong magnetic fields is again a result of (2.15) and the orthogonality of the wave functions within the Landau

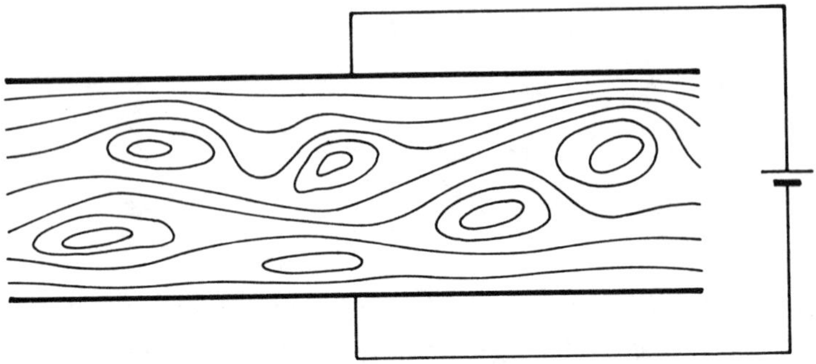

FIG. 4. A schematic illustration of equipotential lines when an electric field is applied across the system edges.

level. It might also be inherent to the peculiar feature of the present system that the energy spectrum is discrete and that each Landau level is completely degenerate in the absence of random potential fluctuations. The problem of the Anderson localization is discussed further in Sec. 3.

2.3 Arbitrary Magnetic Fields

The properties (iii) - (v) described above have been derived in the limit of strong magnetic fields where matrix elements between different Landau levels in (2.4) can be completely neglected. Therefore, there might be corrections due to mixings of different Landau levels. It can easily be shown that such corrections vanish to the lowest order in $(\Gamma/\hbar\omega_c)^2$, where Γ is the broadening of a Landau level. It is not an easy task, however, to extend such kind of a perturbational method to higher order in $(\Gamma/\hbar\omega_c)$. We should remark that, within the single-site approximation, we can show that $\Delta\sigma_{xy} = 0$, i.e. $\sigma_{xy} = -Ne^2/h$, for an arbitrary magnetic field when E_F lies in a gap region where the density of states vanishes.[18] This strongly suggests that the value of the quantized Hall conductivity is not affected by these mixings between Landau levels. Laughlin has presented a plausible argument that this is actually the case.[13] In the following a different argument originally suggested by Takemori[19] will be presented.

We consider an L×L system and impose periodic boundary conditions in the y direction as has been done in the previous section. Electrons are assumed to be free to pass through the left (x=0) and right (x=L)

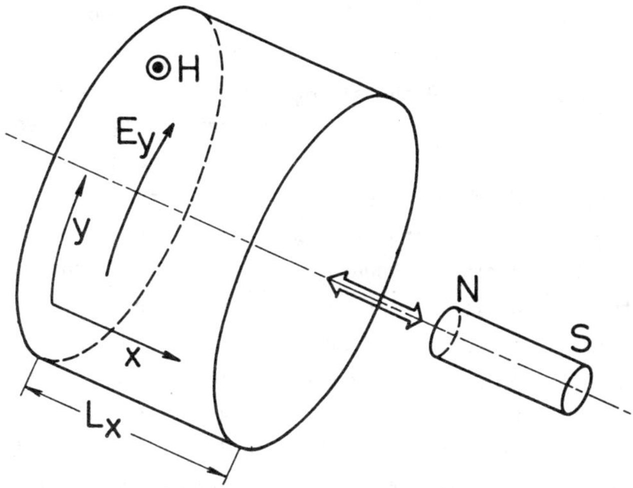

FIG. 5. The configuration used for explaining the quantized Hall effect. The magnetic field is perpendicular to the system consisting of the cylindrical surface. The electric field E_y is generated by moving a magnet along the center of the cylinder.

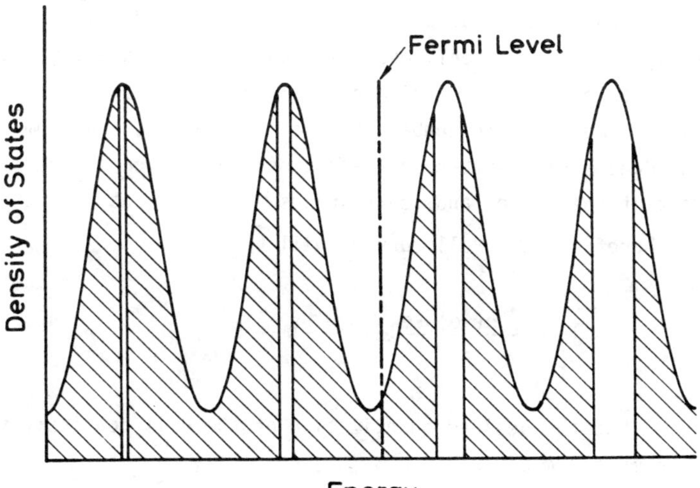

FIG. 6. A schematic illustration of the density of states in magnetic fields. The Fermi level is assumed to be in the localized regime (shaded parts).

edges of the system. (We can also impose periodic boundary conditions in the x direction.) An electric field E_y is introduced by a time-dependent vector potential A_y which is independent of the coordinates x and y. We can imagine the configuration given in Fig. 5. A vector potential A_y independent of the coordinates can usually be eliminated by a change in the phase of the wave function (change in the gauge), i.e., by multiplying the wave function by $\exp(-ieA_y y/\hbar c)$. Because of the periodic boundary conditions in the y direction, however, the elimination is not possible except in case that A_y is the integer multiple of ΔA_y defined by $(2\pi/L)(\hbar c/e)$.

Now we increase A_y very slowly from zero. Let Δt be the time which corresponds to the change of A_y by ΔA_y. Each time A_y reaches the integer multiple of ΔA_y, the Hamiltonian of the system becomes identical with that at t=0 (A_y=0) by the gauge transformation mentioned above. Let us now assume that the Fermi level lies in a nonzero gap in the spectrum of extended states as is shown in Fig. 6. After the time Δt, the wave function or the density matrix of the system becomes identical with the previous one apart from a simple phase difference. This is expected because localized states do not contribute to the current and their occupations are not influenced by the vector potential A_y at all. The only essential change after the time Δt will be that a certain number p of electrons disappear through the left edge of the system and the same number of electrons enter the system through the right edge. This number of electrons p is considered to be independent of the time and corresponds nothing but to the Hall current. Therefore, the Hall current will be

$$j_x = - \frac{p(-e)}{L} \frac{Le}{hc} \frac{\Delta A_y}{\Delta t} = - p \frac{e^2}{h} E_y , \tag{2.16}$$

which leads to the desired quantization of the Hall conductivity into the integer multiple of $-e^2/h$.

The main difference between Laughlin's arguments and the above lies in the origin of the electric field E_y. Laughlin introduced an electrostatic potential instead of the time dependent vector potential. Consequently, extended edge states play an important role in his argument as has been discussed by Halperin.[14)] In contrast, the present argument is more natural and does not require an explicit discussion of the sample edges.

Although the above arguments as well as Laughlin's seem to be sufficiently convincing, these cannot be regarded as a proof of the quantized Hall conductivity in the rigorous sense. Especially, they cannot predict the value of the integer p, and p might be identically zero. As far as the present author knows, no one has succeeded in proving the quantized Hall effect based on the more rigorous Kubo formula for the conductivity.

3. Magnetic Localization - Historical Survey

It has long been known in silicon inversion layers that states are localized near the edges of low-lying Landau levels in strong magnetic fields. This fact can clearly be seen even in the first famous experiments of Fowler, Fang, Howard, and Stiles,[20)] where the conductivity is very small for a considerable range of electron concentrations between the peaks for different spin- and valley-split levels. This problem was left untouched for a long while and then received considerable theoretical and experimental attention.

Kawaji and Wakabayashi[8)] made a careful study of the diagonal conductivity σ_{xx} in strong magnetic fields and observed the vanishing of σ_{xx} in several finite regions of n, indicating the existence of localized electrons at the edges of Landau levels. An example of the experiments is given in Fig. 7. They have shown that the concentration of localized electrons associated with the tail of the Nth Landau level is approximately given by $[2\pi\ell^2(2N+1)]^{-1}$, as is shown in Fig. 8. This roughly means that the electron wave functions become extended when the cyclotron orbit covers the whole space. This interesting experimental result has provoked various theoretical investigations.

The possibility of two-dimensional Wigner crystallization, with and without strong magnetic fields, has been studied theoretically by a number of authors. Tsukada[21)] investigated the properties of the electron crystal in strong magnetic fields, extending a mean field theory of Platzman and Fukuyama[22)] in the absence of a magnetic field. He showed that the N dependence and H dependence of the concentration of localized electrons could be similar to that obtained in the experiments of Kawaji and Wakabayashi. However, the mean field theory

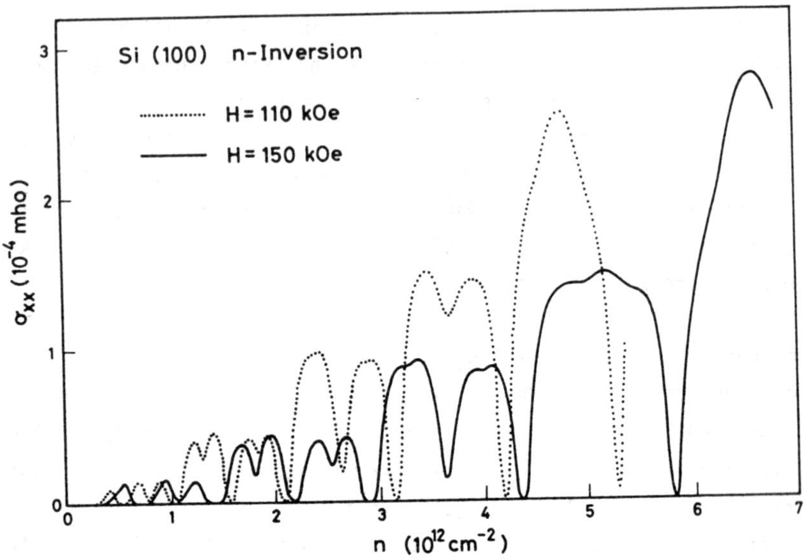

FIG. 7. An example of the observed diagonal conductivity in a Si inversion layer at two different magnetic fields.[8)]

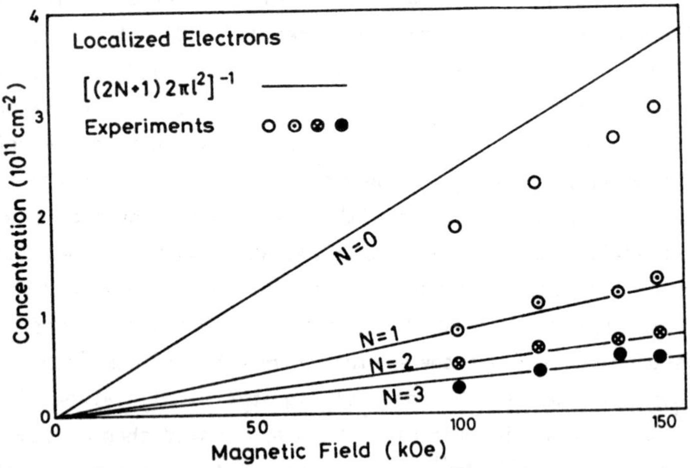

FIG. 8. Concentrations of localized electrons associted with low-lying Landau levels observed by Kawaji and Wakabayashi.[8)]

is known to give a too large critical electron concentration and temperature for the Wigner crystal formation in the absence of a magnetic field. The situation is considered to be the same in strong magnetic fields.

Another possible explanation of the observed localization is the Anderson localization due to potential fluctuations. Tsukada[23] assumed slowly varying potential fluctuations and calculated the density of states in the tail region of Landau levels. In the case of slowly varying potentials the center of the cyclotron motion moves along an equipotential loop. He assumed that the percolation of such loop orbitals around potential minima determined the delocalization of electrons. Assuming simplified circular orbits, he obtained a condition for the localization. In actual inversion layers, however, it is not certain whether the slowly varying limit is applicable or not, because different kinds of scatterers can contribute to scattering and because the short-range scatterer model is known to work quite well.[1] Further, the basic assumption of the percolation of circular loops is highly questionable.

Aoki and Kamimura[24] proposed a condition for the localization by a heuristic argument corresponding to the Ioffe-Regel condition,[25] $k_F \Lambda = 1$, where Λ is the mean free path in the absence of magnetic field. Their argument seems to lead to the existence of a minimum metallic conductivity independent of the Landau-level index. Assuming a value for the minimum metallic conductivity and using the result of σ_{xx} calculated in the self-consistent Born approximation, they derived an expression for the concentration of localized electrons similar to the experimental results.

Actually, there have been various experimental investigations in addition to mentioned above. Kawaji and Wakabayashi[26] measured the activation-type temperature dependence of the diagonal conductivity when the Fermi level lies in the low-energy part of the lowest Landau level. The temperature dependence was shown to be given by a sum of two activated conductivities and to be unable to be explained by the presence of a minimum metallic conductivity. They suggested that this might be explained by a quantum diffusion of Schottky deffects in the Wigner crystal. Kawaji, Wakabayashi, Namiki, and Kusuda found that the temperature dependence was strongly sample-dependent.[27] They suggested the presence of a minimum metallic conductivity in samples with low mobilities.

Nicholas, Stradling, and Tidey measured the temperature dependence of minimum values of σ_{xx} for different H.[28] The experimental results were explained by the existence of the mobility edge and the minimum metallic conductivity. The obtained minimum metallic conductivity turned out to be dependent on the Landau levels and even on samples strongly. Work has been extended to lower temperatures.[29] At extremely low temperatures the dependence of σ_{xx} on the temperature was shown to deviate from exp(-W/T) and obey $\exp(-B/T^{1/3})$. This might suggest that the so-called variable-range hopping is a dominant conduction process. Similar experiments were performed also by Pepper.[30]

At low concentrations, especially where only the lowest Landau level is occupied by electrons, the cyclotron resonance exhibits a peculiar behavior. Kotthaus, Abstreiter, Koch, and Ranvaud observed a rapid decrease of the cyclotron effective mass m_c at low electron concentrations ($n < 4\times10^{11}$ cm^{-2}) in silicon inversion layers for the laser frequency $\hbar\omega = 3.68$ meV.[31] The decrease of m_c was attributed by them to perturbed cyclotron resonance due to potential fluctuations. When a slowly potential fluctuaion V(r) exists, the change of the frequency of the cyclotron motion is given by the curvature of the local potential, i.e., by[32]

$$\delta(h\omega_c) = \frac{1}{2}(\ell\nabla)^2V(r) \quad . \tag{3.1}$$

When electrons occupy the ground Landau level completely, this shift vanishes because the curvature of the potential becomes zero on the average. At lower electron concentrations electrons are considered to be in local potential mimima where $\delta(\hbar\omega_c) > 0$, and the effective cyclotron energy becomes larger or m_c decreases. This simple consideration explains the observed value of the electron concentration where the rapid decrease occurs. A similar shift of the peak can occur for short-range scatterers.[33]

Similar experiments were also performed by Kennedy, Wagner, McCombe, and Tsui.[34] They found a characteristic dependence of m_c on the electron concentration similar to that observed by Kotthaus et al. for a fixed photon energy. They showed, however, that their results depend on the strength of the applied magnetic field. They argued that this was inconsistent with the perturbed cyclotron resonance discussed above and attributed it to some phase change of the two-dimensional electron gas. Wilson, Allen, and Tsui[35] measured the cyclotron resonance at low concentrations with the use of a far-infrared spectrometer. This technique enables a direct observation of the frequency dependent conductivity at a fixed magnetic field. They showed that the resonance peak was shifted to the higher frequency side and the line shape was narrower than that expected from the zero field mobility. Further, the line shape exhibited a peculiar change as a function of the filling factor ν of the lowest Landau level. As a possible explanation they proposed a highly correlated or crystalline ground state.

It is an important problem what the ground state of an interacting electron system is in strong magnetic fields in the absence of random potential fluctuations. Fukuyama, Platzman, and Anderson[36] calculated the free energy within the Hartree-Fock approximation and suggested that the usual phase with a uniform electron density distribution becomes unstable against the formation of charge-density waves at temperatures well above the melting temperature of the classical electron lattice. According to the calculation the instability occurs at $T_c = 0.555\nu(1-\nu)e^2/\ell$, and the wave number of the charge-density wave is independent of the electron concentration given by $1.568/\ell$.

Electronic properties of the charge-density wave ground state have been investigated by various authors. Kuramoto[37] and Kawabata[38], concluded that the charge-density wave state was a kind of zero-gap semiconductor, i.e., there was not an energy gap at the Fermi level, for $\nu=1/2$ (half-filled Landau level). However, a later calculation has shown that the charge-density wave state is essentially an electron lattice and has a large gap at the Fermi level.[39] There have been various investigations on the lattice structure of the charge-density wave.[40,41]

4. Anderson Localization in Strong Magnetic Fields

4.1 Numerical Experiments

One of the most direct methods to investigate the problem of the Anderson localization is a numerical computer experiments. Aoki is the one who first made such numerical study in the two-dimensional system in strong magnetic fields.[42] He calculated wave functions and participation ratios in several sample systems with a finite size containing randomly distributed short-range scatterers. He also calculated matrix elements of current operators. It has been suggested that states are localized near low- and high-energy edges of low-lying Landau levels. In the following results of much more extensive numerical experiments will be presented.

We consider a system with a finite size L×L. Scatterers are distributed randomly and the random potential is given by the sum of individual ones. We shall use periodic boundary conditions for the both x and y directions. To this end, we construct an infinite periodic lattice in which the L×L system constitutes a unit cell and as the basis in the absence of scatterers we choose

$$\psi_{NX}(r) = M^{-1/2} \sum_{m} \phi_{NX+mL}(r) \quad , \tag{4.1}$$

where $\phi_{NX}(r)$ is the usual Landau wave function given by (2.5) and M is the total number of the cells. This $\psi_{NX}(r)$ is periodic in the y direction for $X = j\Delta X$ $(j=1,2,..,N_X)$, where ΔX and N_X have been defined below (2.5). It is periodic also in the x direction except for a phase factor, i.e.,

$$\psi_{NX}(x+L,y) = \exp(-i\frac{Ly}{\ell^2})\psi_{NX}(x,y) \; , \tag{4.2}$$

as is desired. It is clear that $\psi_{NX}(r)$ constitutes an orthonormal basis set.

The matrix element of the potential becomes

$$(NX|V|N'X') = \int dr\, \phi_{NX}^*(r)[V(r)+V(x+L,y)+V(x-L,y)]\phi_{N'X'}(r)$$
$$+ \sum_{\pm} \phi_{NX}^*(r)[V(r)+V(x\mp L,y)]\phi_{N'X\pm L}(r) \quad , \tag{4.3}$$

where couplings among different cells other than between the nearest neighbor cells have been neglected since the system size L is much larger than the cyclotron radius . The magnetic field is assumed to be strong enough for each Landau level to be treated separately. In this

TABLE 1. System parameters used for the numerical experiments.

N_x		36	64	100	144	196	256	324	400
L (ℓ)		15.0	20.0	25.1	30.1	35.1	40.1	45.1	50.1
Number of Samples	N=0	1217	678	328	206	166	144	109	90
	N=1	1217	678	328	206	166	144	112	80
	N=2	1233	680	334	212	170	144	112	80

case the energy levels can be obtained by numerically diagonalizing the matrix $(NX|V|N'X')$ for each N. We assume equal amounts of attractive and repulsive scatterers, which enables us to use a symmetry around the center of each Landau level. That is, the density of states and the diagonal conductivity at T=0 are symmetric around the center E_N of the Landau level N. The Hall conductivity has the symmetry mentioned in Sec. 2.2.

Whether states are localized or extended is determined by a study of the Thouless number g(L) which is defined as the ratio of the shifts $\langle\Delta E\rangle$ of the individual energy levels due to a change in boundary conditions to the level separation $[L^2D(E)]^{-1}$, where D(E) is the density of states per unit area. If states are localized, g(L) is expected to decrease with the increase of the system size L reflecting how the wave function decays with the distance from the localization center. For extended states the energy shifts should be of the order of h/τ, where τ is the time for the electron to diffuse away to the boundary, and g(L) becomes independent of L. This g(L) is the scaling variable in the one-parameter scaling theory of Abrahams et al.[17)]

Licciardello and Thouless[46)] have shown that the conductivity in two dimensions can directly be related to the Thouless number g(L). The same theory can be applied fairly straightforwardly to the present case in strong magnetic fields if we note that the velocity operators v_x and v_y can now be replaced by $\dot{X}$ and $\dot{Y}$, respectively. Under the assumption (i) that the square modulus of velocity matrix element is uncorrelated with the corresponding energy difference and (ii) that couplings between different cells can be well approximated by a tight-binding model for each energy level, we have

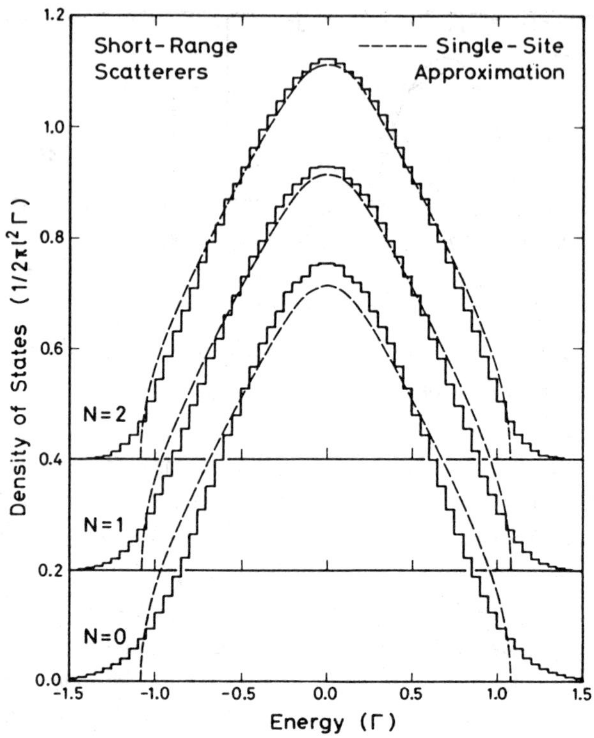

FIG. 9. Histograms of the density of states of the lowest three Landau levels N=0, 1, and 2.

$$\sigma_{xx} = \frac{1}{4}\frac{e^2}{\hbar} g(L) \quad . \tag{4.4}$$

The assumption (ii) should be a reasonable one near the spectral edges because couplings are small compared with spacings between levels. It may not be valid near the center of Landau levels especially with a high level index. The assumption (i) leads to the conclusion that ΔE has a Lorentzian (Cauchy) distribution and $\langle\Delta E\rangle$ should be the geometric mean of the absolute value of individual ΔE.

The energy shifts ΔE are calculated by replacing periodic boundary conditions by antiperiodic conditions in the y direction. This can be achieved by using $X = (j-1/2)\Delta X$ instead of $j\Delta X$. Numerical calculations are performed for several different values of N_X and for the lowest three Landau levels (N = 0, 1, and 2) in case of short-range

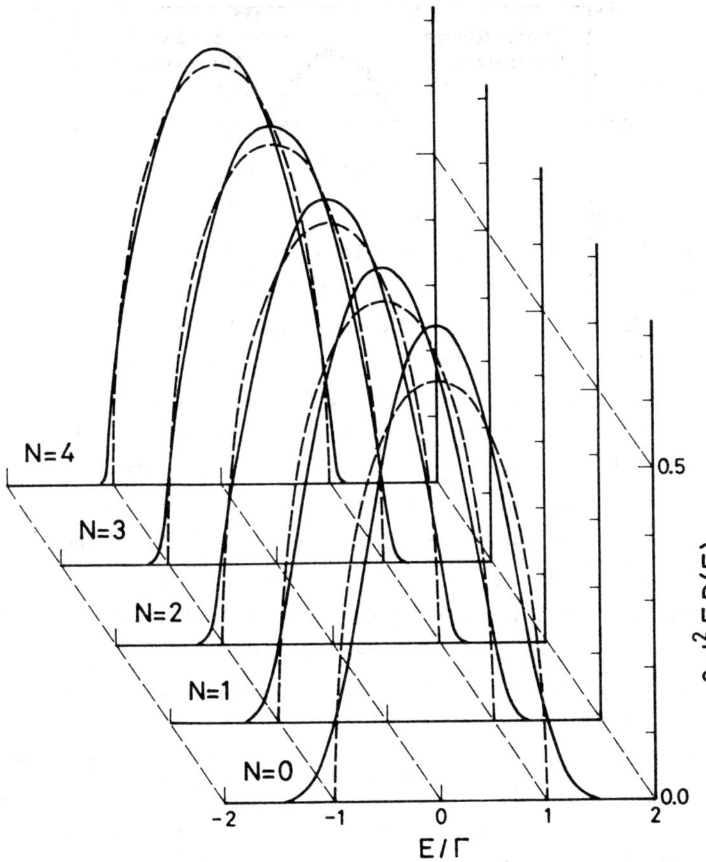

FIG. 10. The density of states of Landau levels calculated theoretically.[48] The dashed lines represent those calculated in the self-consistent Born approximation.

delta-potential scatterers. The values of N_X, corresponding L's, and number of samples are given in Table 1. The actual number of samples is twice as large as that shown in the table owing to the symmetry around the center of the Landau level.

It has previously been shown that for delta-potential scatterers discrete levels which are not affected by scatterers and have the energy E_N appear when their concentration is small.[47] The condition of the appearance is given by $c_i < 1$, where c_i is the average number of scatterers contained in the lowest cyclotron orbit, i.e., $c_i = 2\pi\ell^2 n_i$

with n_i being the concentration of the scatterers. Here, we assume a rather high concentration of scatterers $c_i = 5$. In case of high concentrations effects of a multiple scattering from a single scatterer become unimportant and the single-site approximation gives results close to those in the self-consistent Born approximation.

4.2 Results of Numerical Experiments

Figure 9 shows examples of calculated hitograms of the density of states for N = 0, 1, and 2 together with the results calculated in the single-site approximation. The energy is normalized by the level broadening Γ calculated in the self-consistent Born approximation and its origin is chosen at the center E_N. This broadening Γ is independent of the level index N. In the self-consistent Born approximation, the density of states is given by a semi-elliptic form with the width Γ. The density of states becomes slightly narrower than the dashed line obtained in a single-site approximation. The tails of the density of states becomes smaller and the density of states approaches that calculated in the single-site approximation with increasing level index N. This is in agreement with the result shown in Fig. 10 calculated previously by summing up an infinite perturbation series in an approximate way.[48]

The difficulty of the problem in strong magnetic fields lies in the absence of any expansion parameters like $(2\pi E_F\tau)^{-1}$ in the absence of a field, where E_F is the Fermi energy and τ is the relaxation time. Consequently, almost all diagrams have equal amounts of contributions. The density of states shown in Fig. 10 has been calculated in the approximation that all irreducible diagrams of a given order contribute equally to the self-energy.

Histograms of the Thouless number g(L) are given in Fig. 11, where the logarithm of g(L) for different L's is plotted against energy. Only the lower half ($E < 0$) of Landau levels is shown because of the symmetry. It is clear that states are exponentially localized except in the vicinity of the center of each Landau level.

We can determine the extent of localized wave functions (the localization length), $\alpha(E)^{-1}$, by putting

$$g(L) = g(0)\exp[-\alpha(E)L] \ . \tag{4.5}$$

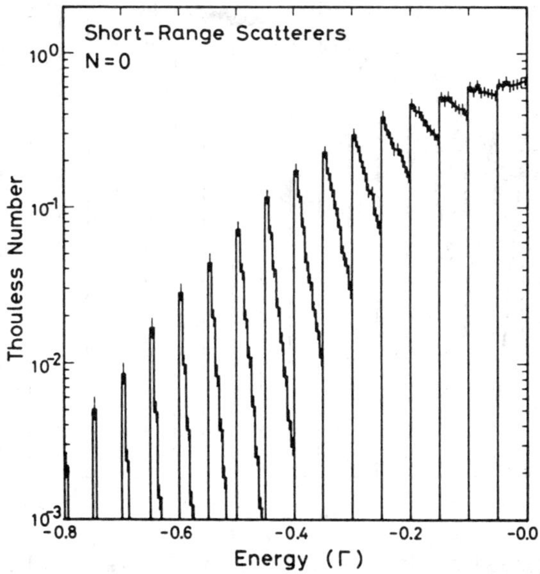
Short-Range Scatterers
N = 0
Thouless Number
Energy (Γ)

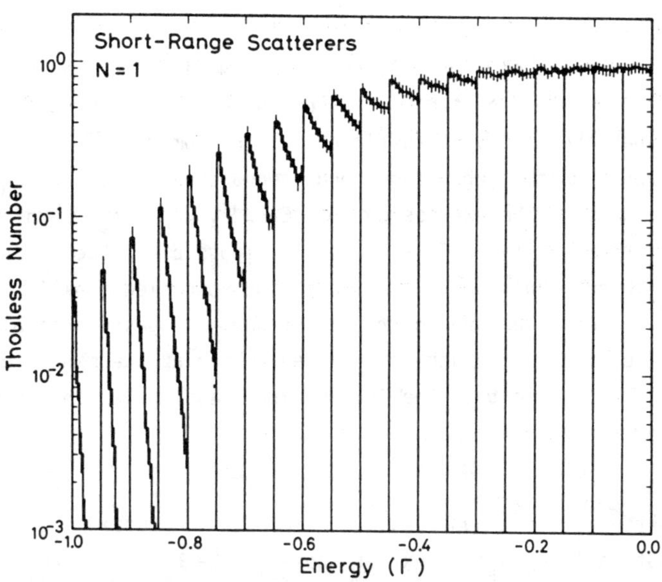
Short-Range Scatterers
N = 1
Thouless Number
Energy (Γ)

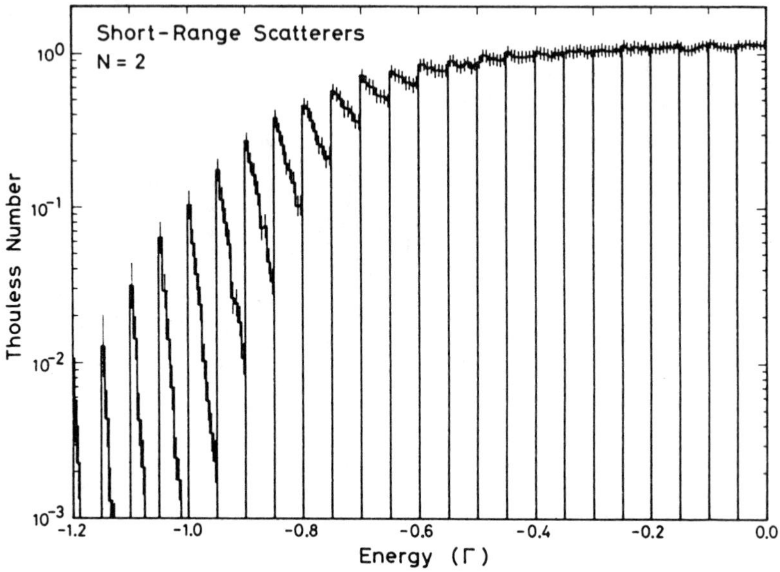

FIG. 11. System-size dependence of the Thouless number for the lowest three Landau levels N = 0, 1, and 2. Within each energy bin the Thouless numbers with increasing system sizes are plotted against the system size.

The resulting α's are plotted against energy in Fig. 12. As is easily seen, the decay rate α of the wave function decreases almost linearly with energy when the energy is sufficiently far from $E = E_N$. However, α approaches zero smoothly when the energy approaches the center E_N. If we fit the results to the expression $\alpha = |E-E_c|^s$, we get s which is much larger than unity. This makes it impossible to determine the position of the critical energy E_c even if the so-called mobility edge is present. Instead, the present result strongly suggests that all the states are localized exponentially except those at the center of the Landau level and also that the inverse localization length α becomes zero smoothly at the center.

4.3 Discussion

The results of the numerical experiments discussed above now suggests that all the states except those at the center of the Landau level are localized exponentially. This conclusion is not in

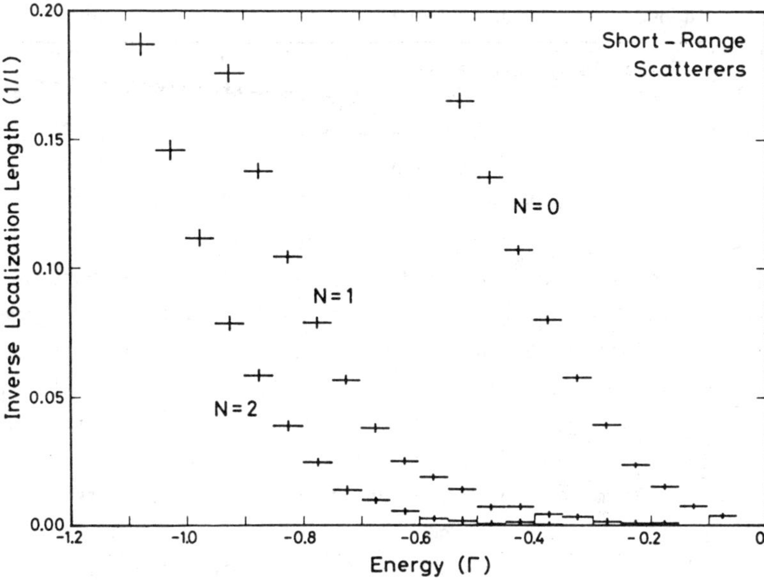

FIG. 12. Calculated inverse localization length as a function of energy for the N=0, 1, and 2 Landau levels.

contradiction with the experimental results of Kawaji and Wakabayashi who showed that the concentration of localized states for the level N was given by $[2\pi\ell^2(2N+1)]^{-1}$ in silicon inversion layers. As is clear in Fig. 12 the extent of localized wave functions is extremely large in the vicinity of the center of the Landau level. Further the energy range corresponding to such a "weakly" localized state becomes wider with the level index N. Experiments are performed at nonzero temperatures. Inelastic electron-electron or electron-phonon scatterings present at nonzero temperatures can easily destroy such weak localization. Apparent mobility edges may be determined by a simple extrapolation of α's in the strong localization regime. This gives the concentrations of localized electrons in good agreement with the experiments of Kawaji and Wakabayashi.

The result of the numerical experiments does not contradict the conclusion (v) obtained in Sec. 2 in connection with the quantized Hall effect, either. One might think that the argument given below (2.15) leads to a peculiar conclusion that an electron in the extended state at the center of the Landau level must move with an infinitely large velocity in order to compensate the decrease of the Hall current due to the localization of other levels. This is not necessarily the case, however, because the conclusion that all the states are localized is applicable only in the absence of an electric field. The presence of an applied electric field can delocalize some of weakly localized states in the vicinity of the center of the Landau level and create current-carrying states having a nonzero measure.

Quite recently, Ono[49)] reached the same conclusion as that of the numerical experiments by extending a self-consistent diagramatic technique developed by Vollhardt and Wölfle[50)] to the case in strong magnetic fields. He calculated the decay rate for the lowest Landau level N=0, which goes to zero continuously at the center of the Landau level. However, the energy dependence of α is different from the numerical result mentioned above. Ono's calculation predicts that $\alpha(E) \propto \exp(-\Gamma^2/E^2)$ near E=0. The present result shows much more gradual decrease of α when E approaches E=0. It is certainly impossible to give a quantitatively correct answer by such diagramatic perturbation expansions, since every diagram gives almost same amounts of contributions as has already been noted previously.

So far only the case of short-range scatterers has been considered. Actually, however, the nature of the localization depends drastically on the range of the potential of scatterers. In case that the potential range is much larger than the radius of the cyclotron orbit, the electron motion becomes almost classical and an electron moves along an equipotential line, as has already been noted several times. Consequently, the electron wave function is determined essentially by the distribution of the local potential energy. This fact can be clearly demonstrated in Fig. 13, which gives an example of the local potential energy and corresponding eigen wave functions. In these figures, scatterers are assumed to have a Gaussian potential with a range $d = 2\ell$. Numerical simulation has also been performed for the case of such long-range scatterers. Within the present model, we can also calculate physical quantities like the Hall conductivity, the dynamical conductivity, and the line shape of the cyclotron resonance. Those will be discussed elsewhere.[45)]

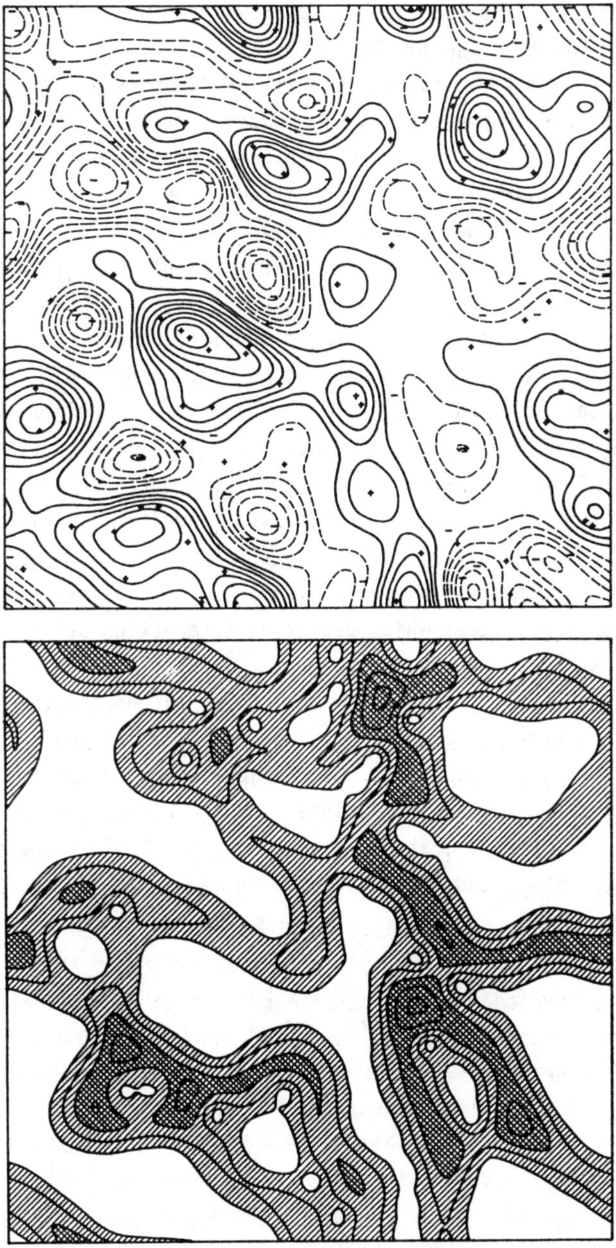

FIG. 13. An example of equipotential lines and absolute value of the wave function of a state near the center of the lowest Landau level in case of long-range scatterers.

5. Problems to be Answered in Future

5.1 Critical Magnetic Field?

The one-parameter scaling theory of Abrahams et al.[17)] predicts that states are exponentially localized in two dimensions in the absence of a magnetic field. This conclusion has subsequently been confirmed by a renormalization-group calculation based on the nonlinear σ model.[51)] When a weak magnetic field is applied to the system, the magnetic field tends to increase the extent of localized wave functions. This effect gives rise to the negative magnetoresistance widely observed in varius quasi-two-dimensional systems.[52,53)] This is consistent with the conclusion reached in Sec. 2 that all the states cannot be localized in strong magnetic fields. It is believed, however, that the magnetic field does not destroy the localization as long as it is sufficiently weak. Therefore, the important question arises what the strength of the critical field is. If such a critical field exists, it is determined by the level broadening Γ in case of high concentrations of short-range scatterers since Γ is, apart from the level index N, the only relevant parameter which describes interaction between different Landau levels.

One way to study this problem will be to extend the theory of Aoki and Ando[12)] to weaker magnetic fields and see effects of mixings between different Landau levels. Corrections lowest order in $\Gamma/\hbar\omega_c$ have already been shown to vanish. Higher order effects have not been considered yet. However, it is likely that the Hall conductivity is singular at $\Gamma/\hbar\omega_c$ and the expansion in terms of $\Gamma/\hbar\omega_c$ does not have much meaning at all.

As has been discussed in the last paragraph of Sec. 2.2, the presence of delocalized states in the Landau level in the strong-field limit seems to be linked to (2.15) and the orthogonality of the wave functions within the Landau level. Therefore, it is highly likely that the critical field is infinite, i.e., mixings among different Landau levels make the delocalized states at the center of the Landau level a weakly localized state. If this is the case, both the diagonal conductivity σ_{xx} and the Hall conductivity σ_{xy} vanishes eventually when both the applied electric field and the temperature are decreased further and further. In any case, future investigations (both experimental and theoretical) in this direction are highly expected.

5.2 Quantization into Noninteger Multiple of $-e^2/h$?

Quite recently, Tsui, Störmer, and Gossard found a very striking result in the two-dimensional system made on the GaAs-$Al_xGa_{1-x}As$ heterojunction interface.[54] According to the experiments, the Hall conductivity is quantized into $-e^2/3h$ and $-2e^2/3h$ when the Fermi level lies in the lowest (N=0) Landau level. They speculated that this is due to the presence of triangular electron lattices. The argument presented in Sec. 2.3 does not allow such quantization into noninteger multiple of $-e^2/h$. Although the discussion does not explicitly take into account mutual interactions among electrons, it is expected to be valid even in the presence of electron-electron interactions. Therefore, the question arises whether the formation of the electron lattice allows the Hall conductivity to take noninteger times $-e^2/h$ and why the noninteger is 1/3 or 2/3. Thouless, Kohmoto, Nightingale, and den Nijs[55] recently considered effects of periodic potentials on the quantized Hall effect. The Hall conductivity takes various different values depending on the magnetic flux contained in the single period of the potential, each being an integer multiple of $-e^2/h$ in agreement of the discussion in Sec. 2.3.

5.3 Lowest Landau Level in Si Inversion Layers

As is shown in Figs. 3 and 7, the lowest Landau level splits into four levels. This is due to the removal of the spin degeneracy and also of the valley degeneracy. The diagonal conductivity becomes quite small when the Fermi level lies in the lowest of the four N=0 Landau levels. Correspondingly the absolute value of the Hall conductivity becomes much smaller than that of $-e^2/h$ when the Fermi level lies between the lowest two N=0 levels. This means that states are completely localized in this level. However, when the Fermi level lies between the spin-splitted levels (the second and third N=0 levels from below), the Hall conductivity becomes quantized into $-2e^2/h$. This fact suggests that at this electron concentration, i.e., at $n = 2/2\pi\ell^2$, some of the states associated with the lowest level which are previously localized have now become extended states.

There are two possibilities in explaining this interesting behavior. Firstly, the strength of random potential fluctuations is dependent on the electron concentration. At low electron concentrations potential fluctuations are considered to be extremely strong because they are not screened by electrons in the inversion layer. Therefore, the strong-field condition is not satisfied and states are completely localized according to the scaling theory of Abrahams et al.[17] At higher concentrations, however, such strong potential fluctuations are screened out by electrons and the strong-field condition is satisfied. The second possibility is of course the crystallization into electron lattices due to strong electron-electron interactions at low concentrations. With increasing concentration the electron-electron Coulomb interaction becomes less important and the electron lattice eventually melts, resulting in the delocalization. At present, it is not possible to distinguish between the two possibilities. It is highly likely that the two effects are almost equally important and the system can be regarded as a Wigner glass.

6. Summary and Conclusion

We have discussed various problems related to the localization in the two-dimensional system in strong magnetic fields. The most dramatic effect of the localization may be the quantization of the Hall conductivity σ_{xy} into an integer multiple of $-e^2/h$. This quantized Hall effect has been the subject of Sec. 2. Important exact properties of the Hall conductivity worked out by Aoki and Ando have been discussed based on the Kubo formula. An argument has been presented which suggests that the Hall conductivity should be an integer multiple of $-e^2/h$ when the Fermi level lies in the localized regime of the energy spectrum.

The theory of Aoki and Ando[12] shows that all the states cannot be localized in the strong-field limit where mixings between different Landau levels are completely neglected. This is quite in contrast to the result in weak magnetic fields, where all the states are localized in two dimensions. The results of numerical simulations given in Sec. 4 shows, however, that only states at the center of the Landau level are extended and all the others are exponentially localized. That is, there is no mobility edge in the Landau level. There remains a serious problem on whether this conclusion is valid only in the strong-field limit and whether all states become localized if mixings between Landau levels are present.

In actual two-dimensional inversion layers, mutual electron-

electron-interactions can play an important role. This, for example, might lead to a crystallization of the electron gas. The interplay of random potential fluctuations and electron-electron interactions is a remaining important theoretical problem. This problem has briefly been discussed in Secs. 3 and 5.

Acknowledgments

I would like to thank Dr. H. Aoki for valuable correspondences. The numerical computation has been performed with the aid of HITAC M200H/M280H at the University of Tokyo and FACOM M200 at the University of Tsukuba. A part of this work has been supported by the Grant-in-Aid for Fundamental Research from the Ministry of Education.

References

1) T. Ando, A.B. Fowler, and F. Stern, Rev. Mod. Phys. 54, 437 (1982).
2) K. von Klitzing, G. Dorda, and M. Pepper, Phys. Rev. Lett. 45, 494 (1980).
3) S. Kawaji and J. Wakabayashi, in Physics in High Magnetic Fields, edited by S. Chikazumi and M. Miura (Springer, Berlin, 1981), p. 284.
4) D.C. Tsui and A.C. Gossard, Appl. Phys. Lett. 38, 550 (1981).
5) K. Yoshihiro, J. Kinoshita, K. Inagaki, C. Yamanouchi, J. Moriyama, and S. Kawaji, Surf. Sci. 113, 16 (1982); J. Phys. Soc. Jpn. 51, 5 (1982).
6) T. Ando and Y. Uemura, J. Phys. Soc. Jpn. 36, 959 (1974).
7) T. Ando, Y. Matsumoto, Y. Uemura, M. Kobayashi, and K.F. Komatsubara, J. Phys. Soc. Jpn. 32, 859 (1972).
8) S. Kawaji and J. Wakabayashi, Surf. Sci. 58, 238 (1976).
9) T. Ando, Y. Matsumoto, and Y. Uemura, J. Phys. Soc. Jpn. 39, 279 (1975).
10) S. Kawaji, T. Igarashi, and J. Wakabayashi, Prog. Theor. Phys. Suppl. 57, 176 (1975).
11) R.E. Prange, Phys. Rev. B 23, 4802 (1981).
12) H. Aoki and T. Ando, Solid State Commun. 38, 1079 (1981).
13) R.B. Laughlin, Phys. Rev. B 23, 5632 (1981).
14) B.I. Halperin, Phys. Rev. B 25, 5849 (1982).
15) D.J. Thouless, J. Phys. C 14, 3475 (1981).
16) R. Kubo, S.J. Miyake, and N. Hashitsume, Solid State Phys. edited by F. Seitz and D. Turnball (Academic Press, New York, 1965), Vol. 17, p. 269.
17) E. Abrahams, P.W. Anderson, D.C. Licciardello, and T.V. Ramakrishnan, Phys. Rev. Lett. 42, 673 (1979).
18) T. Ando, unpublished.
19) T. Takemori, private communication.
20) A.B. Fowler, F.F. Fang, W.E. Howard, and P.J. Stiles, Phys. Rev. Lett. 16, 901 (1966).
21) M. Tsukada, J. Phys. Soc. Jpn. 42, 391 (1977).
22) P.M. Platzman and H. Fukuyama, Phys. Rev. B 10, 3150 (1974).

23) M. Tsukada, J. Phys. Soc. Jpn. 41, 1466 (1976).
24) H. Aoki and H. Kamimura, Solid State Commun. 21, 45 (1977).
25) N.F. Mott and E.A. Davis, Electronic Processes in Non-Crystalline Materials, 2nd ed. (Clarendon, Oxford, 1979).
26) S. Kawaji and J. Wakabayashi, Solid State Commun. 22, 87 (1977).
27) S. Kawaji, J. Wakabayashi, M. Namiki, and K. Kusuda, Surf. Sci. 73, 121 (1978).
28) R.J. Nicholas, R.A. Stradling, and R.J. Tidey, Solid State Commun. 23, 341 (1977).
29) R.J. Nicholas, E. Kress-Rogers, F. Kucher, M. Pepper, J.C. Portal, and R.A. Stradling, Surf. Sci. 98, 283 (1980).
30) M. Pepper, Philos. Mag. B 37, 83 (1978).
31) J.P. Kotthaus, G. Abstreiter, J.F. Koch, and R. Ranvaud, Phys. Rev. Lett. 34, 151 (1974).
32) S.J. Miyake, J. Phys. Soc. Jpn. 20, 412 (1965).
33) T. Ando, J. Phys. Soc. Jpn. 38, 989 (1975).
34) T.A. Kennedy, R.J. Wagner, B.D. McCombe, and D.C. Tsui, Phys. Rev. Lett. 35, 1031 (1975) R.J. Wagner, T.A. Kennedy, B.D. McCombe, and D.C. Tsui, Phys. Rev. B 22, 945 (1980).
35) B.A. Wilson, S.J. Allen, Jr. and D.C. Tsui, Phys. Rev. Lett. 44, 479 (1980).
36) H. Fukuyama, P.M. Platzman, and P.W. Anderson, Phys. Rev. B 19, 5211 (1979).
37) Y. Kuramoto, J. Phys. Soc. Jpn. 44, 1035 (1978).
38) A. Kawabata, Solid State Commun. 28, 547 (1978).
39) D. Yoshioka and H. Fukuyama, J. Phys. Soc. Jpn. 47, 394 (1979).
40) R.R. Gerhardts and Y. Kuramoto, Z. Phys. B 44, 301 (1981).
41) Y. Kuramoto, to be published.
42) H. Aoki, J. Phys. C 10, 2583 (1977); 11, 3823 (1978).
43) T. Ando, Surf. Sci. 113, 182 (1982).
44) T. Ando, in Anderson Localization, edited by Y. Nagaoka and H. Fukuyama (Springer-Verlag, Berlin, 1982), p. 176.
45) T. Ando, to be published.
46) D.C. Licciardello and D.J. Thouless, J. Phys. C 8, 4157 (1975); 11, 925 (1978).
47) T. Ando, J. Phys. Soc. Jpn. 36, 1521 (1974).
48) T. Ando, J. Phys. Soc. Jpn. 37, 622 (1974).
49) Y. Ono, J. Phys. Soc. Jpn. 51, 2055 (1982).
50) D. Vollhardt and P. Wölfle, Phys. Rev. B 22, 4666 (1980).
51) See, for example, F.J. Wegner, in Anderson Localization, edited by Y. Nagaoka and H. Fukuyama (Springer-Verlag, Berlin, 1982), p. 8.
52) S. Kawaguchi and S. Kawaji, J. Phys. Soc. Jpn. 48, 699 (1980).
53) S. Hikami, A.I. Larkin, and Y. Nagaoka, Prog. Theor. Phys. 63, 707 (1980).
54) D.C. Tsui, H.L. Störmer, and A.C. Gossard, Phys. Rev. Lett. 48, 1559 (1982).
55) D.J. Thouless, M. Kohmoto, M.P. Nightingale, and M. den Nijs, Phys. Rev. Lett. 49, 405 (1982).

LOCALIZATION IN TWO-DIMENSIONAL SYSTEMS: SILICON MOS INVERSION LAYERS IN WEAK AND STRONG MAGNETIC FIELDS

S. Kawaji

Department of Physics, Gakushuin University
Mejiro, Toshima-ku, Tokyo, 171 Japan

Electrons in inversion layers in Si-MOSFETs (Metal-Oxide-Semiconductor Field-Effect-Transistors) behave as a two dimensional electron system. The two-dimensional nature of the system in a random potential appears in two ways in response to magnetic field: (1) angular dependent negative magnetoresistance in weak magnetic fields and (2) the quantum Hall effect in strong magnetic fields. Experimental researches on these phenomena in our group are reviewed.

1. Introduction

In recent years, rapid progress has been made in understanding of surfaces and interfaces of solid materials. The progress in the researches of solid surfaces or vacuum-solid interfaces have mainly been made in understanding of atomic and electronic structures of surfaces. On the contrary, the progress in the researches of solid-solid interfaces such as the silicon-silicon dioxide interface in MOSFETs (Metal-Oxide-Semiconductor Field-Effect Transistors), the III-V compound semiconductor heterojunction interfaces and super-lattice structures have mainly been made in observations of new electronic phenomena which can not be searched in bulk materials.

The silicon-silicon dioxide system is technologically most important interface in the present age of electronics because Si-MOSFETs are vital elements to the computer hardware. From the viewpoint of science, silicon inversion layers in high quality Si-MOSFETs produced by advanced silicon technology have been developing experimental bases in the physics of two-dimensional electron systems[1].

A schematic structure and an energy level diagram of an n-channel Si-MOSFET on a (001) surface of a p-type Si substrate are shown in Fig. 1. The surface density N_s of electron number in the inversion layer can be controlled by changing the gate voltage V_G. Because of

a

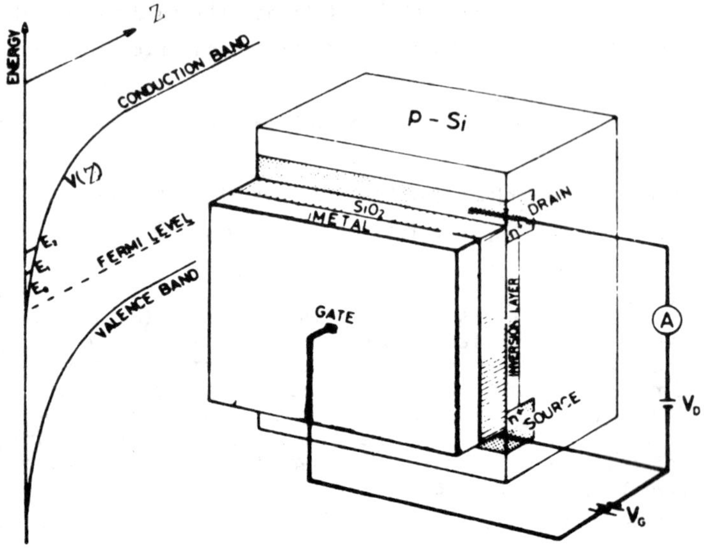

b

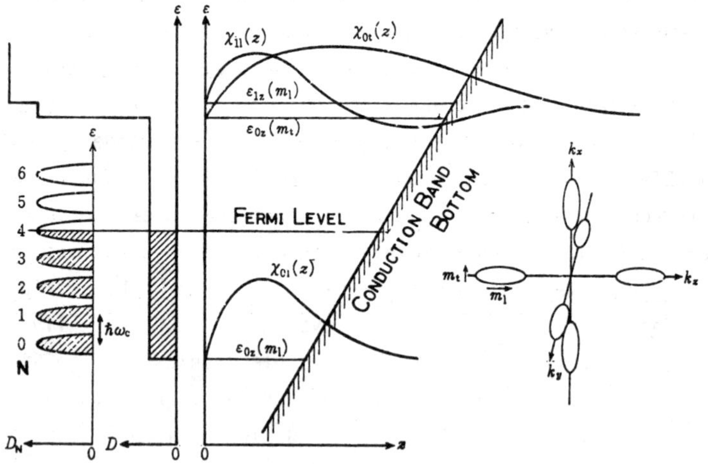

FIG. 1 (a). Schematic illustration of an n-channel Si-MOSFET and a rough sketch of the energy diagram of the inversion layer. (b) Energy level diagram of an n-channel Si(001) inversion layer. D and D_N are the density of states without and with application of a strong magnetic field.

the thin ($\sim 10^{-7}$ cm) inverted triangular potential bounded by the silicon dioxide band gap at the silicon-silicon dioxide interface and the bottom of the silicon conduction band with high gradient normal to the interface (z-direction) generated by the applied electric field, the electron motion in the z-direction is quantized so that the electrons distribute in two-dimensional (2D) subbands with continuums for motion in the plane parallel to the interface whose density of state is shown as D in Fig. 1 (b). At low temperature T and low N_s, electrons occupy the ground subband only so that electrons behave completely as a 2D electron gas. When a strong mangetic field is spplied along a direction normal to the surface, the continuum of electron energy levels in 2D motion coalesces into a series of perfectly quantized Landau levels shown as D_N in Fig. 1 (b)[2]. The 2D nature of electrons and the controllability of the electron density in a single device by direct electrical means make the inversion layer in Si-MOSFETs most suitable for studying the fundamental properties including many electron effects of the 2D electron gas.

The 2D nature of such systems appears clearly in the transport properties of the systems in random potentials in response to magnetic fields at low temperatures. In the following, results of our experimental researches in two problems will be described: (1) the negative magnetoresistance in weak magnetic fields in the weakly localized regime of electrical conduction in Si-MOSFETs and (2) the electron localization in Landau levels observed in the transverse and the Hall conductivity, σ_{xx} and σ_{xy}, and the quantum Hall effect of Si-MOSFETs in strong magnetic fields.

2. Negative Magnetoresistance in Anderson Localization in Silicon Inversion Layers

2-1 Historical and Theoretical Backgrounds

Before 1979, several observations of negative magnetoresistance in 2D electron systems in semiconductor inversion and accumulation layers had been reported[3∿10]. Uemura and Matsumoto's theory[11] based on an idea of the change in the scattering probability of 2D electrons defected in a direction normal to the surface due to the Lorentz force has explained the negative magnetoresistance observed in tangential magnetic field by Tansal, Fowler and Cotellessa[4] and Sakaki and Sugano[5]. Other observations had been anomalous phenomena. The negative magnetoresistance observed in InAs inversion layers[3] and ZnO accumulation layers[10] had shown very weak anisotropy on the field direction which is favourable to a localized spin scattering model. However, Kawaguchi, Kitahara and Kawaji's observations in Si(111)-MOS inversion layers[8] and silicon inversion layers produced by Cs adsorption[9] had shown clearly that the negative magnetoresistance depends only on the field component perpendicular to the surface as shown in Fig. 2. This anisotropy shows that the negative magnetoresistance is caused by the 2D motion of conduction electrons.

In 1979, Abrahams, Anderson, Licciardello and Ramakrishnan[12] have developed a scaling theory of the Anderson localization. The theory has shown that the localization in 2D random systems has particular features that electrons are always localized at absolute zero even if the impurity scattering is very weak and the conductance decreases with sample size L, logarithmically in high conductance region and exponentially in low coductance region. Anderson, Abrahams and Ramakrishnan[13] have shown based on Thouless's Argument[14] that the sample size L can be replaced by $\sqrt{D\tau_\varepsilon}$ at finite temperatures, where D is the diffusion coefficient and τ_ε is the inelastic scattering time. Thus, the log L dependence of conductance can be replaced by log T dependence due to the temperature dependence of the inelastic scattering time $\tau_\varepsilon \propto T^{-p}$. Their theory has explained log T dependence of conductance in metallic films observed by Doran and Osheroff[15]. Bishop, Tsui and Dynes[16] have explained their observations of log T dependence of conductance in Si-MOSFETs by the scaling theory of Anderson localization.

On the basis of the scaling theory of the Anderson localization, Hikami, Larkin and Nagaoka[17] and Altshuler, Kmel'nizkii, Larkin and Lee[18], have developed the theory of negative magnetoresistance in the weakly localized regime (WLR) ($2\pi\varepsilon_F\tau/\hbar \gg 1$) of 2D systems. In WLR the quantum interference between states with e^{ikr} and e^{-ikr} causes spacially localized states which produce the correction term σ' to the classical Drude conductivity σ_0. The quantum correction is sensitive to perturbations which destroy time reversal symmetry. Therefore, an external magnetic field applied perpendicularly to the

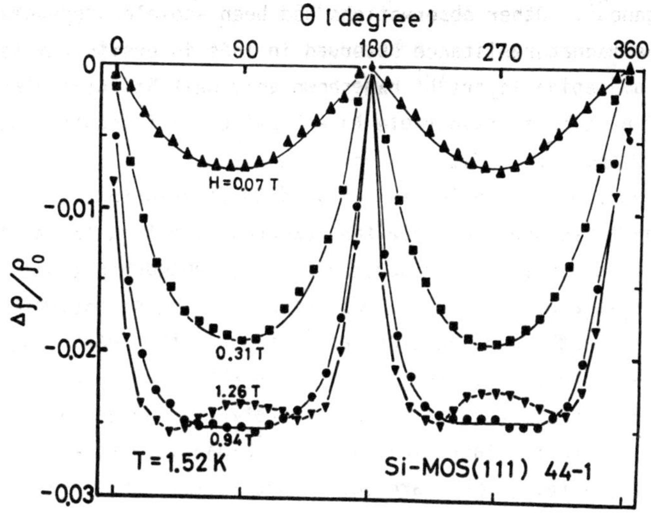

FIG. 2a. Angular dependence of transverse magnetoresistance of a Si(111) MOS inversion layer at $N_S=5\times10^{12}$ cm^{-2}, where θ=0° at B // surface.[8)]

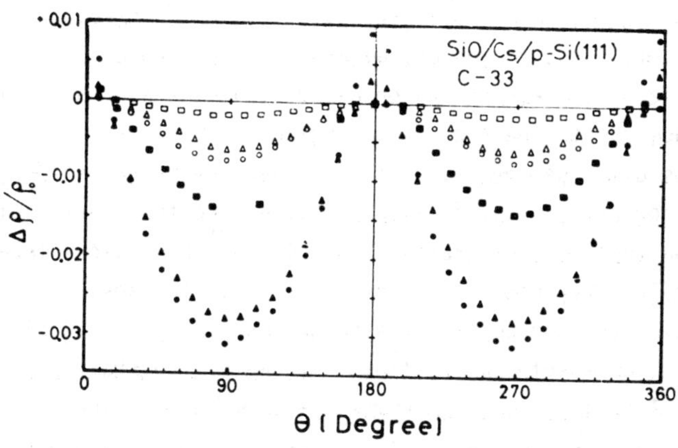

FIG. 2b. Angular dependence of transverse magneto-resistance of a cesiated Si(111) inversion layer at temperatures 4.2K (□), 2.3K(Δ) and 1.7K(○), where θ=0° at B // surface. The open symbols indicate the field B=0.31T and full symbols indicates B=1.25T.[9)]

2D systems suppresses the quantum correction to the conductivity due to the weak localization and produces negative magnetoresistance. The theory has explained recent experiments in Si-MOSFETs by Kawaguchi and Kawaji[19,20,21], Uren, Davies, Kaveh and Pepper[22], Wheeler[23] and Bishop, Dynes and Tsui[24].

The quantum correction to the conductivity in WLR of 2D systems has logarithmic dependence on temperatures, electric fields as well as on the magnetic fields. However, in these dependences, the negative magnetoresistance is probably most useful to investigate microscopic parameters in the localization theory such as the inelastic scattering time.

The quantum correction to the conductivity due to the localization depends also on the magnetic scattering[17] and the spin-orbit interaction[17, 25] though they are not important in Si-MOS inversion layers. In n-channel Si-MOS inversion layers, the intervalley scattering plays a dominant role in the quantum correction due to the localization[26].

In addition to the localization theoty, Altshuler, Aronov and Lee[27] have pointed out that the mutual coulomb interactions of electrons in 2D systems produce another correction to the Drude conductivity which depends also on log T. Fukuyama[28] has pointed out that the correction due to the mutual interactions causes also a magnetoresistance which depends on log H and reduces the negative magnetoresistance due to the localization theory.

As is discussed by Kaveh et al[29] and Kaveh and Mott[30], the quantum correction to the conductivity due to the localization may depends on the electric field. Tsuzuki[31,32] has calculated the electric field effect on the localization in the absence and the presence of the magnetic fields. The possiblity of the elctric field effect on the quantum correction is, however, controversial at present between Altshuler, Aronov and Kmel'nizkii[33] and Tsuzuki[34].

In view of the theoretical progress in the quantum correction to the Drude conductivity due to the localization effect and the interaction effect described above, detailed analysis of negative magnetoresistance experiments in Si-MOS inverison layers affords us fruitful informations on electron localization and interactions in 2D electron systems[35,36]. In this situation, the controllability of the electron number density in a single sample by direct electrical means makes the

2D systems in Si-MOSFETs the best for studying those fundamental properties of the 2D electron systems.

2-2 Fundamental Properties of Negative Magnetoresistance in Silicon MOS Inversion Layers

Following Hikami, Larkin and Nagaoka[17], the change in conducitivity with magnetic field B applied perpendicularly to the 2D system having impurity Coulomb scattering only is written as

$$\Delta\sigma_B = \sigma(B) - \sigma(0) = -\frac{n_v\alpha e^2}{2\pi^2\hbar}\left[\psi\left(\frac{1}{2}+\frac{1}{a\tau}\right) - \psi\left(\frac{1}{2}+\frac{1}{a\tau_\varepsilon}\right) - \ln\frac{\tau_\varepsilon}{\tau}\right] \quad (1)$$

where n_v is the valley degeneracy α is a constant prefactor, τ is the relaxation time due to normal impurity scattering, τ_ε is the inelastic scattering time and $a = 4DeB/\hbar$ where D is the diffusion coefficient. This formula is applicable for $a\tau < 1$ because $(a\tau)^{-1} \sim N_0$ is the cut-off parameter in the summation of $\Sigma\{1/[(N+1/2)a+1/\tau_\varepsilon]\}$. Here we note that $a\tau$ is given by

$$a\tau = \frac{4\varepsilon_F\omega_c\tau^2}{\hbar} = \frac{4\pi\hbar\sigma_0\omega_c\tau}{n_v e^2} = \frac{\sigma_0(\text{mho})\mu B}{1.93\times10^{-5} n_v} \quad (2)$$

where μ is the electron mobility. Equation (1) gives also the correct result for the saturation value of the magnetoconductivity at high field limit,

$$\Delta\sigma_B(\text{sat}) = \frac{n_v\alpha e^2}{2\pi^2\hbar}\ln\frac{\tau_\varepsilon}{\tau} \quad (3)$$

In the analysis of experiments[19,20,21], we consider that the constant prefactor α includes effects of the intervalley scattering[26] and the mutual Coulombinteractions[28] which are not taken into account by Hikami et al[17]. Therefore, in our analysis, τ_ε and α have been determined to reproduce most well experimental magnetorconducitivity by eq. (1). Wheeler[23] and Bishop, Dynes and Tsui[24] have employed $n_v\alpha = 1$ neglecting the mutual interactions and assuming the strong intervalley scattering limit. However, the best fit of eq.(1) to experiments has been obtained in our analysis by adjusting both $n_v\alpha$ and τ_ε.

Figure 3 shows changes in the conducitivity with magnetic field in n-channel (001) Si-MOSFETs with the peak electron mobilities at 4.2K

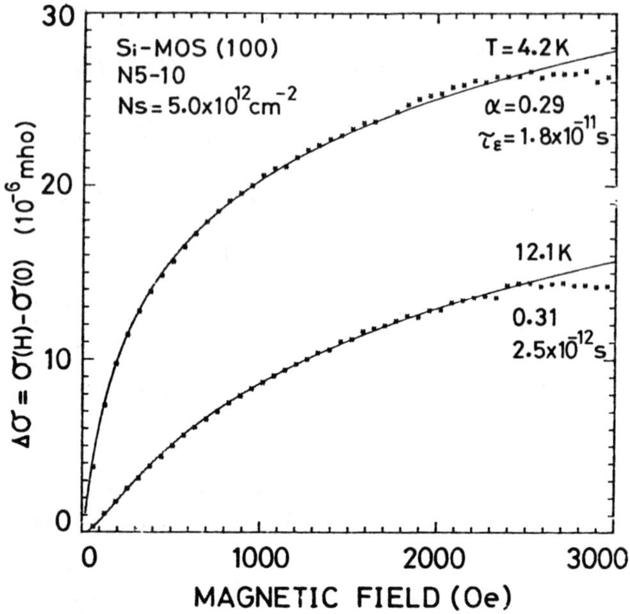

FIG. 3a. Magnetoconducitivy in an n-channel(100) Si MOSFET with μ (peak at 4.2K) = 13000 $cm^2/v \cdot s$. The experimental points are fitted to the expression (1) with an approximation $\psi(1/2 + 1/\tau a) \approx \ell n(1/\tau a)$.[19)]

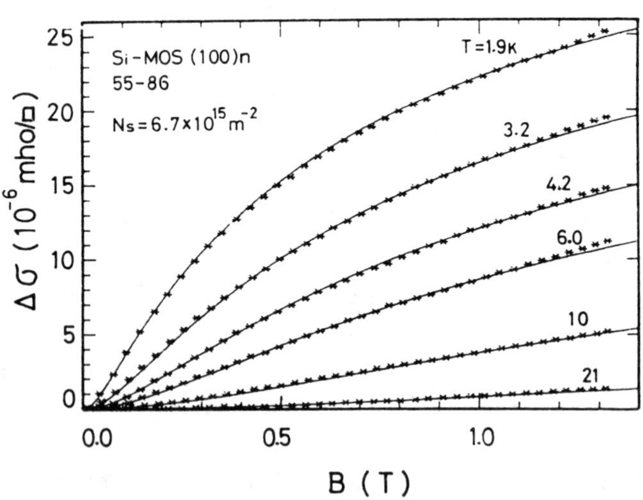

FIG. 3b. Magnetoconductivity in an n-channel (100) Si MOSFET with μ (peak at 4.2K) = 3500 $cm^2/v \cdot s$. The experimental points are fitted to the expression (1).

of 13000 $cm^2/v \cdot s$ (a)[19] and of 3500 $cm^2/v \cdot s$(b)[21]. The experimental points (crosses) are fitted to an approximation of eq.(1) in Fig. 3a and to eq.(1) in Fig. 3b expressed by solid lines. In Fig. 3a, $a\tau$ is about 3.3 at the field of 3 kOe. In the calculations of eq.(1) in Fig. 3b, $a\tau$ is kept less than 0.6. It is interesting to note here that the experiments can be well fitted to eq.(1) for $a\tau > 1$. Variation of the fitting parameters $n_v\alpha$ and τ_ε is less than 10% for $a\tau \leq 4$.

Temperature dependence and electron concentration dependence of inelastic scattering time are shown in Fig. 4 and Fig. 5, respectively. The vertical broken line in Fig. 5 indicates the electron concentration N_s for $2\pi\varepsilon_F\tau/\hbar = \pi$ or $k_F\Lambda = 1$ where k_F and Λ are the Fermi wave number and the mean free path of electrons, respectively. The open and solid squares in Fig. 5 are deduced from magnetoconductivity experiments by use of eq.(1) for $n_v = 1$. Open and solid circles are similarly deduced from the same data for the same N_s for $n_v = 2$. The inelastic scattering time data presented in Fig. 4 are limited to the region where the non-linear electric field effect predicted by Tsuzuki[31,32] is excluded. The data in Fig. 5 satisfy the same condition.

Following Landau argument, τ_ε due to the electron-electron process in the pure metal is written as[14,37]

$$\tau_\varepsilon(\text{el-el, pure}) \simeq \hbar \frac{\varepsilon_F}{(k_B T)^2} \tag{4}$$

Equation (4) gives $\tau_\varepsilon \simeq 5.6 \times 10^{-10}$ s for n-channel (001) Si MOS inversion layers ($n_v = 2$, $m^* = 0.19\ m_o$) at $N_s = 1 \times 10^{12}\ cm^{-2}$ at T=1K.

Recently Abrahams, Anderson, Lee and Ramakrishnan[38] have calculated τ_ε due to the electron-electron process in the dirty 2D system. Their results is written as

$$\tau_\varepsilon(\text{el-el, dirty}) = - \frac{\tau\ \varepsilon_F}{k_B T \ln(T/T_1)} \tag{5}$$

where τ is the normal elastic impurity scattering time, $k_B T_1 = \hbar D \varepsilon^2/e^4$ and $\varepsilon = \hbar D q_s^2$ where q_s is the inverse screening length. The numerical value of T_1 is given as $T_1(K) = 32.6 \times (\sigma_o(\text{mho})/1.23\times10^{-5})^3\ (7.7/\bar{K})^2$ $(m^*/0.19m_o)\ (2/n_v)$.

Calculation of the inelastic scattering time due to the electron-

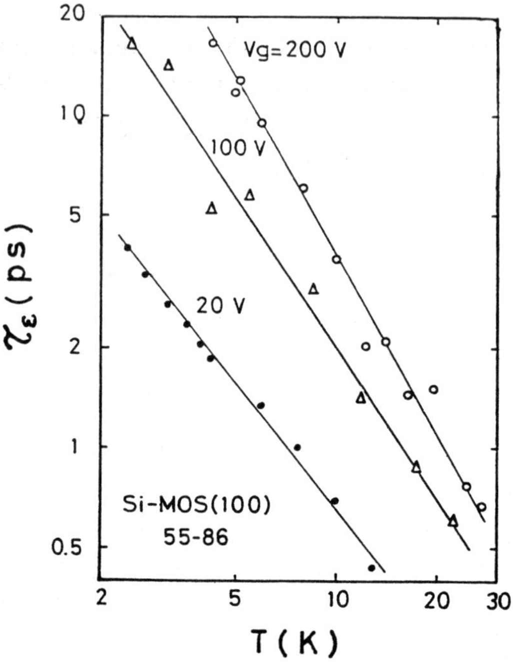

FIG. 4. Temperature dependence of inelastic scattering time τ_ε for $E_F\tau/\hbar$ = 0.4 (V_G=20V), 9.5(V_G=100V) and 14.5(200V).[21)]

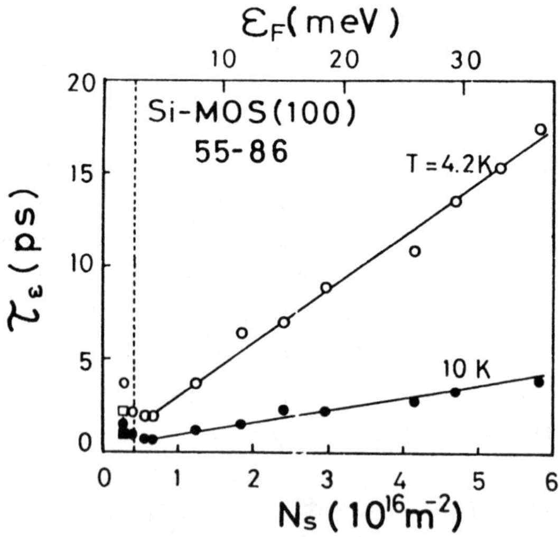

FIG. 5. Electron concentration dependence of inelastic scattering time τ_ε. $E_F\tau/\hbar$ = 0.5 at the vertical broken line.[21)] The open and solid squares are deduced from the magnetoconductivity by use of the expression (1) for n_v = 1, and circles for n_v=2.

phonon process which is applicable to 2D systems in semiconductor inversion layers has not yet been performed. Shinba, Nakamura, Fukuchi and Sakata[39] have calculated the energy relaxation time from the 2D electron system to the phonon system in Si(001) and (111) n-channel inversion layers. Their results show that

$$\tau_{relax}(\text{el-ph}) \propto \frac{\varepsilon_F^{1/2}}{T^2} \quad \text{for } T_e \sim T \tag{6a}$$

$$\propto \frac{\varepsilon_F^{1/2}}{T^3} \quad \text{for } T_e \gg T \tag{6b}$$

at temperature T below 7 K where T_e is the electron temperature. The inelastic scattering time due to the electron-phonon process τ_ε(el-ph) has probably the same dependences on ε_F and T to those in τ_{relax}(el-ph).

Results of N_s-dependence of τ_ε in Fig. 5 show that $\tau_\varepsilon \propto N_s$ which is in accordance with the theoretical results of the electron-electron process in eq(4) and eq(5). However, the exponent p of the temperature dependence of $\tau_\varepsilon \propto T^{-p}$ in experiments lies in the range of $1 < p < 2$ as shown in Fig. 6; p = 1.75 at V_G = 200 V ($N_s = 5.8 \times 10^{12} cm^{-2}$) and p = 1.3 at V_G = 20 V ($N_s = 6.8 \times 10^{11}$ cm^{-2}). The tendency in the N_s-dependence of p is not unreasonable when one considers that the system approaches to the pure metal limit at high N_s and to the dirty limit at low N_s. The inelastic scattering time calculated by eq(4) for the pure metal limit is larger than experiments but not so far from them; $\tau_\varepsilon(\text{theor})/\tau_\varepsilon(\text{exp}) \simeq 6 \sim 15$ for V_G = 200 V in the temperature range in Fig. 4. This result suggests that some other scattering processes exist to reduce the τ_ε and P. However, the inelastic scattering time calculated by eq(5) for V_G = 20 V is one order of magnitude smaller than the experiments whose temperature dependence is close to the results of eq(5) derived for the dirty 2D system.

The constant prefactor α shows an interesting behaviour against the electron concentration N_s as shown in Fig. 7. Fukuyama[35,36] has accounted for the behaviour at higher N_s region by the intervalley impurity scattering effect on the prefactor of the logarithmic correction to the Drude conductivity.

Fukuyama[36] has shown that the intervalley scattering plays an important role in many valley systems and we have

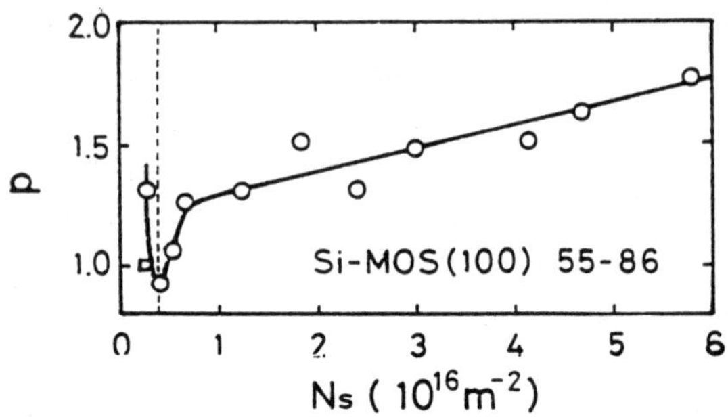

FIG. 6. Electron concentration dependence of exponent P in $\tau_\varepsilon \propto T^{-p}$. $E_F\tau/\hbar = 0.5$ at the vertical broken line.[21] A square is deduced from magnetoconductivity by use of the expression (1) for $n_v = 1$, and circles for $n_v = 2$.

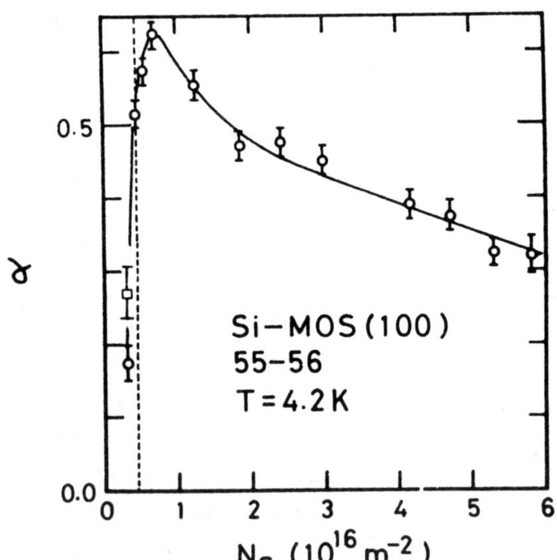

FIG. 7. Electron concentration dependence of prefactor α at 4.2K. $E_F\tau/\hbar = 0.5$ at the vertical broken line.[21] A square is deduced from magnetoconductivity by use of the expression (1) for $n_v = 1$, and circles for $n_v = 2$.

$$\alpha = 1 + F(\frac{1}{2} - n_v)/n_v \qquad \text{for } \tau' \gg \tau_\varepsilon \qquad (7a)$$

$$\alpha = \frac{1}{n_v}[1 - \frac{F}{2n_v}] \qquad \text{for } \tau' \ll \tau_\varepsilon \qquad (7b)$$

where F is the average of the matrix element of scattering in RPA screened Coulomb interactions between two states on the Fermi surface introduced by Altshuler et al[18] and τ' is intervalley impurity scattering time. The quantity F is a function of $2k_F/q_s$ where q_s is inverse screening length and $F \to 1$ as $2k_F/q_s \to 0$ and $F \to 0$ as $2k_F/q_s \to \infty$. In the strong screening limit, α is unity when $\tau' \gg \tau_\varepsilon$ and α is $1/n_v$ when $\tau' \ll \tau_\varepsilon$. The inelastic scattering time τ_ε is proportional to N_s as shown in Fig.5. Therefore, when F = 0, the prefactor α is expected to crosses over from $\alpha \simeq 1$ for $\tau' \gg \tau_\varepsilon$ to smaller value $\alpha \sim 0.5$ for $\tau' \ll \tau_\varepsilon$ with increasing N_s. Actually, following Fukuyama's calculation, $F \simeq 0.8$ at $N_s = 6 \times 10^{12}$ cm^{-2}, then we have from eq(7b) $\alpha \simeq 0.4$. This is close to experimental value of α at $N_s = 6 \times 10^{12}$ cm^{-2} in Fig. 7.

The effect of the intervalley impurity scattering appears also in temperature dependence of α[21]. The prefactor α starts to increase with increasing temperature at 8 K and 22 K for V_G = 20 V ($N_s = 6.8 \times 10^{11}$ cm^{-2}) and V_G = 200 V($N_s = 5.8 \times 10^{12}$ cm^{-2}), respectively. The inelastic scattering time τ_ε is about 0.8×10^{-12} s at these critical temperatures for each gate voltage as shown in Fig. 4. The intervalley impurity scattering time τ' is independent of temperature and is probably close to the value of $\tau_\varepsilon = 0.8 \times 10^{-12}$ s. This is consistent with τ' in the calculation of the structure of the conductivity associated with the minigap by Ando[40].

Fukuyama[35] has pointed out that the sharp drop of α with decreasing N_s at low N_s region ($N_s < 5 \times 10^{11}$ cm^{-2}) in Fig. 7 is consistent with the removal of the valley degeneracy, i.e. $n_v = 2 \to 1$ at low N_s. Following Ohkawa and Uemura's theory[38] of valley splitting due to tunneling effect in k-space through Γ point between two valleys, the bare valley splitting at $N_s = 5 \times 10^{11}$ cm^{-2} is about 0.15 meV which is smaller than the valley splitting with can explain $n_v = 2 \to 1$ at this N_s. Bloss, Sham and Vinter[41] have shown that intravalley exchange and correlation induce a first order phase transition and $n_v = 2 \to 1$

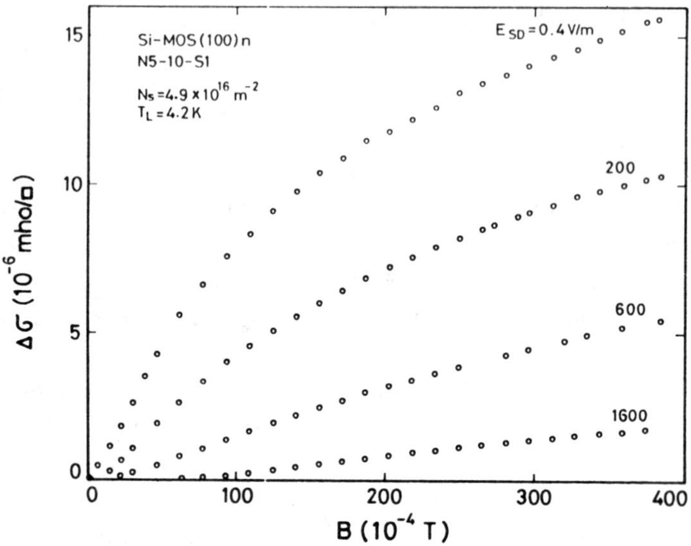

FIG. 8. Magnetoconductivity of an n-channel (100) Si MOSFET at various source-drian fields.[42,43]

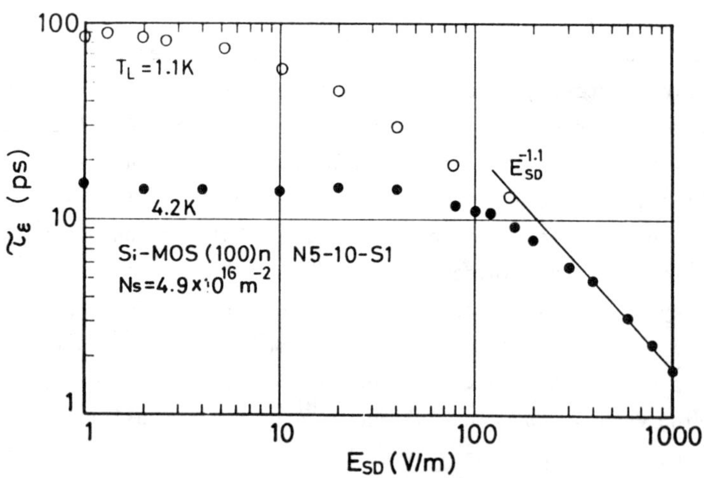

FIG. 9. Inelastic scattering time versus source-drain field in a n-channel (100) Si MOSFET at lattice temperatures of 4.2K and 1.1K.[42]

occurs roughly below $N_s = 3 \times 10^{11}$ cm^{-2}. Experimental magnetoconductivity data have been analyzed assuming that $n_v = 1$ and results are shwon by squares in Figs. 5, 6 and 7. They appear more reasonable than the values of τ_ε, P and α which are obtained by taking $n_v = 2$ at the same electron concentration N_s.

2-3 Hot-Electron Effect in Negative Magnetoresistance in Silicon MOS Inversion Layers and Electron-Phonon Coupling

Kawaguchi and Kawaji[42,43] have applied the negative magnetoresistance to measurements of the temperature of electrons in silicon MOS inversion layers at high electric fields to evaluate the deformation potential constants in the electron-phonon interaction. As is discussed in the former section, the inelastic scattering time τ_ε is determined by electron-electron interaction. Therefore, we can measure the temperature of 2D electron systems in thermally non-equilibrium state to the lattice by τ_ε deduced from the negative magnetoresistance experiments. As is shown in Fig. 4, the inelastic scattering time depends strongly on temperature of electrons. Therefore, the inelastic scattering time serves probably as the best thermometer for electron temperature.

Figure 8 shows magnetoconductivities of (001) Si MOS inversion layers with peak mobility of 11,000 $cm^2/v\cdot s$ at 4.2 K at various source-drain field E_{SD} at lattice temperature T_L of 4.2 K. Similar magnetoconductivity measurments have been performed at $T_L = 1.1$ K. Changes in inelastic scattering time deduced from the magnetoconductivity by use of eq(1) with the source-drain field are shown in Fig. 9. Experiments were performed at $N_s = 4.9 \times 10^{12}$ cm^{-2} where the electron mobility is 5,300 $cm^2/V\cdot s$. The inelastic scattering time data in Fig. 9 show that the electric field E_{SD} lies within the linear region of quantum correction σ' in view of Tsuzuki's results[31,32].

The electron temperature evaluated from temperature dependence of τ_ε at $E_{SD} = 1$ v/m changes against the source-drain field at two lattice temperatures as shown in Fig. 10. At 4.2 K, the change in the electron temperature for the change in the source-drain field of 40V/m $\sim$ 800 V/m is about 14 K. For this change in the electron temperature, the conductivity changes only 1.5 %[43] whereas the inelastic scattering time changes almost one order of magnitude as shown in Fig. 9.

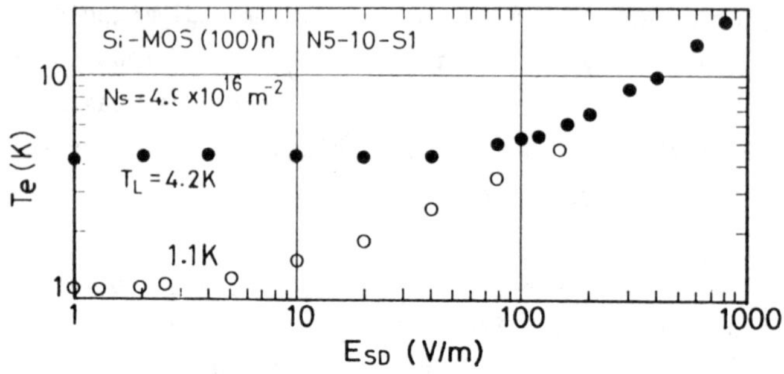

FIG. 10. Source-drain field dependence of electron temperature in a n-channel (100) Si MOSFET at lattice temperature of 4.2K and 1.1K.[42,43]

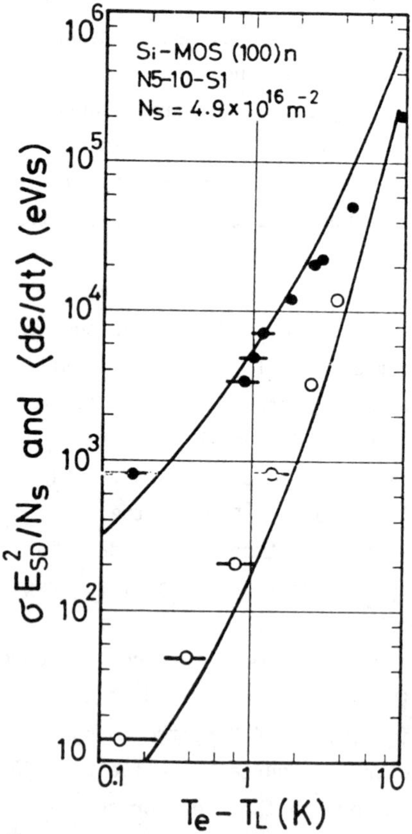

FIG. 11. T_e(electron temperature)-T_L(lattice temperature) versus energy gain per electron from the field ($\sigma E_{DS}{}^2/N_S$) (solid circles at T_L=4.2K and open circles at 1.1K) and energy loss per electron calculated by Shinba and Nakamura (private communication).[42,43]

Electrical conductivity data σ_0 of the sample for different source-drain fields at two lattice temperature of 4.2 K and 1.1 K lie on the same line of the conducitivity versus lattice temperature when we plot the data against the electron temperature[43]. This result shows that the electron mobility depends not on the lattice temperature but on the electron temperature in the present experiment. Therefore, the contribution of the surfon scattering[44] to the electron mobility is weak at the temperature at least lower than 20 K. We note here that the conductivity changes logarithmically with temperature at temperatures lower than 4 K. The coefficient of log T term is in good agreement with the results of Bishop et al[16].

In stationary state at high source-drain field E_{SD}, the rate of energy gain per electron from field, $\sigma_0 E_{SD}^2/N_s$, is equal to the rate of energy loss to the lattice system, $-\langle d\varepsilon/dt \rangle$. The energy loss per electron from the electron system to the lattice by electron-surfon scattering has been calculated by Shinba et al[39] by using the deformation potential constants of bulk silicon, Ξ_u = 12 eV, D = Ξ_d/Ξ_u = 0.67. Calculated values of $-\langle d\varepsilon/dt \rangle$ (solid lines) and experimental data of $\sigma_0 E_{SD}^2/N_s$ (solid and open circles) are shown in Fig. 11 as a function of $(T_e - T_L)$. Good agreement between theoretical results and experiments shows that the deformation potential constants in the inversion layer are almost same to the bulk values.

The result described above, i.e. the deformation potential constants in inversion layers are the same to the bulk values, leads a very important conlclusion that the electron mobility in Si-MOSFETs limited by electron-phonon scattering is much higher than the highest value obtained at present stage of silicon MOS thechnology. The highest mobility limited by the electron-surfon scattering and the surface roughness scattering[45] only is expected to be about 1,000,000 $cm^2/v{\cdot}s$ at 4.2 K and 30,000 $cm^2/v{\cdot}s$ at 77 K.

3. Localization in Strong Magnetic Fields and the Quantum Hall Effect in Silicon MOS Inversion Layers

3-1 Localization in Landau Levels

In n-channel MOS inversion layers on Si(001) surfaces at sufficiently low temperatures and in sufficiently strong magnetic fields, it is possible to realized an extreme-quantum-limit condition (k_BT <

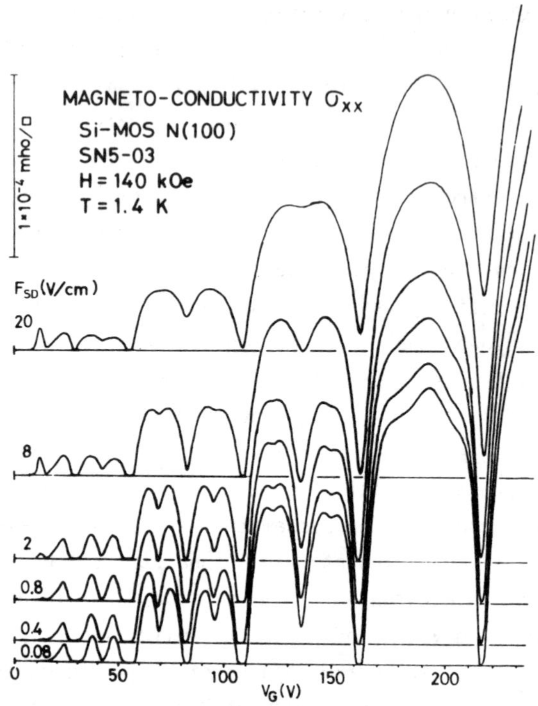

FIG. 12. Transverse conductivity σ_{xx} of an n-channel Si(100) MOSFET versus gate voltage V_G in 14T at 1.4K for various source drain fields F_{SD}[48].

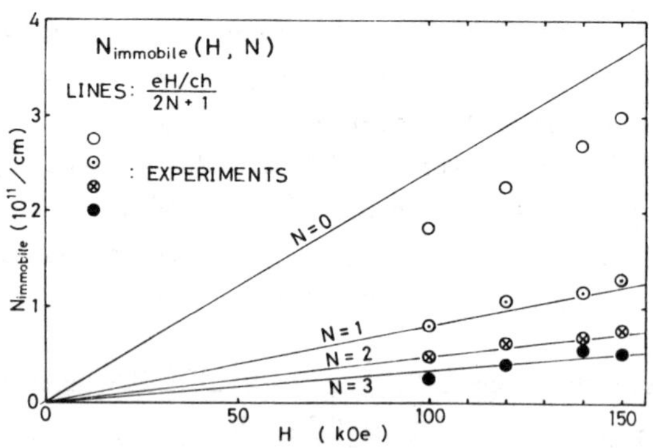

FIG. 13. Concentration of immobile electrons associated with the higher edge of the N-1th Landau level and the lower edge of the Nth Landau level versus magnetic field for different Landau level index N.[48]

$\Gamma < \hbar\omega_c$), where Γ the broadening of the Landau levels and ω_c the cyclotron frequency[2,46]. In such a system, there exist gap regions in the density of states of electrons between the boundaries of each Landau level. When random potentials are incorporated in the system, localized states are expected to exist at higher and lower edges of each Landau level. If the strength of the random potential is weaker than the mutual Coulomb interaction of electrons, the strong magnetic field is expected to enhance correlated states of electrons, i.e., the Wigner solid or the CDW state.

The first experiment of Shubnikov-de Haas effect in Si-MOSFETs by Fowler, Fang, Howard and Stiles[47] has shown the existence of gap regions in the gate voltages where the transverse conductivity σ_{xx} vanishes.

Kawaji and Wakabayashi have performed careful experiments on the finite gap regions in the gate voltages for the vanishing σ_{xx} with various source-drain fields and investigated the magnetic field dependence and the Landau level index dependence of the concentration of immoble electrons[48]. The transverse conductivity σ_{xx} in a Si-MOSFET in 14 T at 1.4 K for various source-drain electric fields is shown in Fig. 12. The magnetic field dependence of the width of the gate voltage for the vanishing conductivity region ($\sigma_{xx} < 10^{-9}$ mho) has shown a result that the sum of the concentration of immobile or localized electrons associated with the higher edge of the (N-1)th Landau level and the lower edge of the N-th Landau level where N is the Landau level index, is approximately given by $[2\pi\ell^2(2N+1)]^{-1}$ as shown in Fig. 13 where ℓ is the radius of the ground Landau orbit. This result means that the electron wave functions becomes extended or pinned correlated states become free when the whole area of the inversion layer is covered by cyclotron orbits with radious of $(2N+1)^{1/2}\ell$.

There exist several experimental and theoretical investigations which suggest possibilities of the Anderson localization due to random potentials and the Wigner crystallization or the CDW states in 2D systems in strong magnetic fields[59]. Recent numerical investigations of the Anderson localization by Ando[49,50,59] have given the concentration of localized electrons which is in good agreement with our experiments.

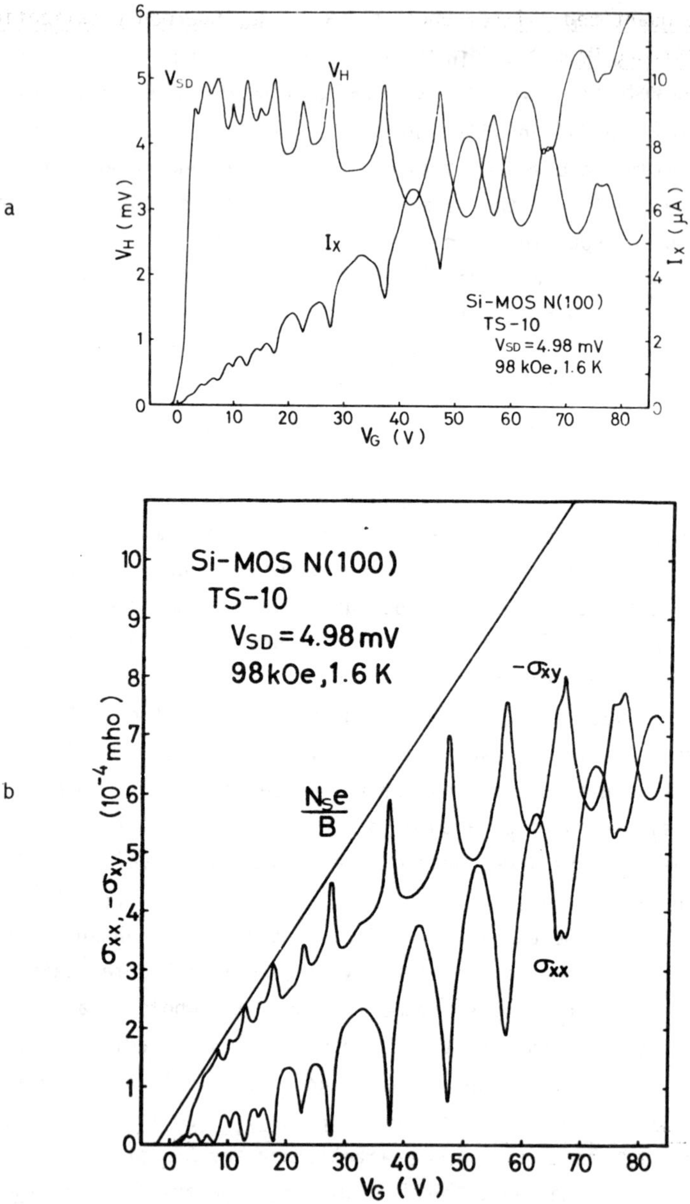

FIG. 14 (a). Channel current I_X and Hall voltage V_H versus gate voltage V_G in an n-channel Si(100) MOSFET with the width to length ratio of 3. (b) Transverse conductivity σ_{xx} and Hall conductivity σ_{xy} from the data in (a).[52,53]

3-2 The Quantized Hall Conductivity and the Anderson Localization in Strong Magnetic Fields

Ando, Matsumoto and Uemura[51] have developed a theory of Hall effect in 2D systems in strong magnetic fields. Their results at T = 0 are summarized as followings:

(1) The Hall conductivity σ_{xy} is given by

$$\sigma_{xy} = - \frac{N_s e}{B} + \frac{\Gamma}{\hbar\omega_c} \sigma_{xx} \quad . \tag{8}$$

(2) The Hall conductivity is not affected by the presence of impurities when each Landau level is completely filled and the Hall conductivity is given by

$$\sigma_{xy} = - \frac{Ne^2}{h} \tag{9}$$

when the Fermi level lies in the energy gap between the (N-1)th and Nth Landau levels.

(3) In the case of impurity bands are separated from each Landau level, the Hall conductivity is given by eq.(9) when the Fermi level lies in any spectral gaps lying between the(N-1)th and Nth main Landau level, i.e. when the Fermi level lies in gaps between two impurity bands or between an impurity band and the main Landau level.

Igarashi, Wakabayashi and Kawaji[52,53] have confirmed experimentally Ando et al's result (2) by Hall voltage measurements for wide samples in 9.8 T at 1.6 K as shown in Fig. 14. The difference between $N_s e/B$ and $-\sigma_{xy}$ in Fig. 14 is much larger than $\sigma_{xx}\Gamma/\hbar\omega_c$ where Γ is calculated by Ando and Uemura's results of self-consistent Born approximation[54].

By employing the Hall current method, Wakabayashi and Kawaji[55] have first confirmed experimentally Ando et al's result (1) for the Hall conductivity.

In 1980, Kawaji and Wakabayashi[56] have reported precise behaviours of the transverse and Hall conductivity and their temperature dependence observed in 15 T at temperatures between 1.5 K and 12 K. Figure 15 shows an example of our results at 1.5 K which clearly

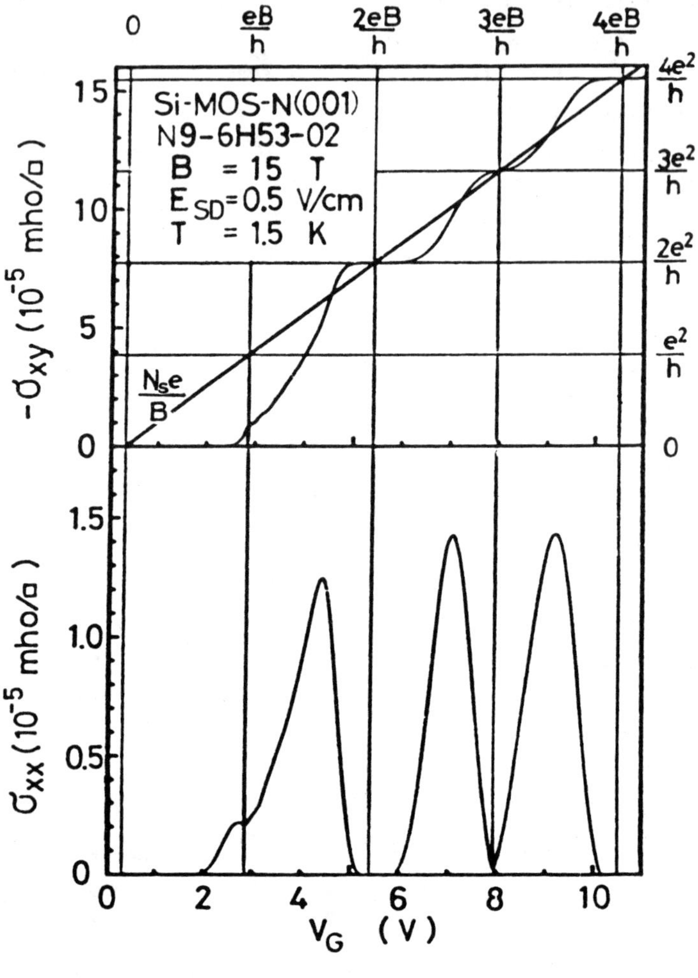

FIG. 15. Hall conductivity σ_{xy} and transverse conductivity σ_{xx} versus gate voltage V_G in the N = 0 Landau level in an n-channel Si(100) MOSFET measured by Hall current method in 15T at 1.5 K. Plateaus in Hall conductivity at $2e^2/h$ and $4e^2/h$ are observed for $\sigma_{xx} = 0$ where the Fermi level lies in the localized states near the gap between (0↑-) and (0↓+) Landau levels and in the localized states near the gap between (0↓-) and (1↑+) Ladau level, respectively. Reference 56 contains temperature dependence of the similar data.

demonstrates the quantized steps in the Hall conductivity for $2e^2/h$ in the spin gap between the (0↑-) level the and (0↓+) level and $4e^2/h$ in the Landau gap between the (0↓-) level and the (1↑+) level.

The quantized Hall conductivity in Fig. 15 demonstrates the characteristic feature of the electron-hole symmetry; i.e., near the spin gap, the Hall conductivity becomes $2e^2/h$ when the Fermi level lies in the localizated states in the upper edge of the (0↑-) level even if the (0↑-) level is not completely filled or when the holes in the (0↑-) level are localized as well as the case where the electrons in the (0↓+) level are localized or the Fermi level lies in the localized states in the lower edge of the (0↓+) level.

The quantized Hall steps and the electron -hole symmetry in the Anderson localization in Landau levels in 2D systems have been expected by Ando et al's theory of the Hall effect in 2D systems for long. One of the important conclusion reached by Ando et al is their results (3) which is obtained for the case where the impurity bands are separated from the main Landau level. The single-site approximation which Ando et al employed has lead the impurity bands separated from the main Landau level. However, localized states exist actually near the edges of each Landau level instead of the impurity bands. Therefore, it is easy to expect the appearance of the quantized Hall plateaus and the electron-hole symmetry in the localization in Landau levels in 2D systems. For example,[57] let us imagine a situation where there are many impurity levels separated each other with very small gaps near the edges of each Landau level and these impurity levels have no electrical conduction. Then we can expect the quantized Hall plateaus and the electron-hole symmetry from Ando et al's result(3).

von Klitzing, Dorda and Pepper[58] have independently observed the quantized Hall steps in the Hall resistance measurements. Although their observations lacked the electron-hole symmetry, their results demonstrated that the Hall resistance for the gap between the (0↓-) level and the (1↑+) level agrees with the recommended value of $h/4e^2$ in high precision of 3 ppm. On the basis of this fact, they proposed a new method of the determination of the fine structure constant $\alpha = e^2/\hbar c$ by combination of the quantized Hall resistance and the light velocity c. Their proposal has attracted many metrologists and physicists in broad areas to the phenomenon of the quantized Hall

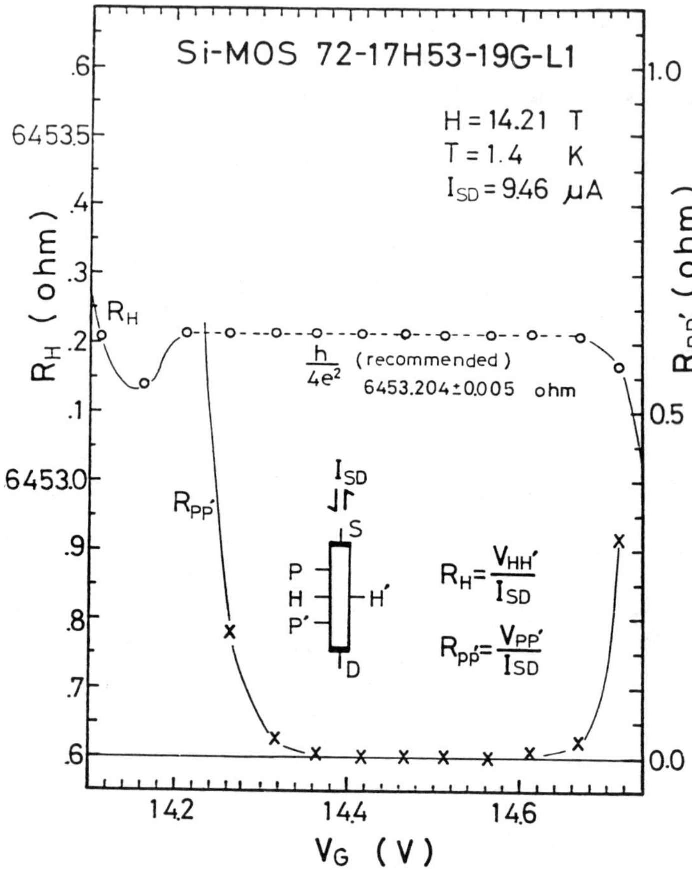

FIG. 16. Results of high precision measurements of Hall resistance R_H and transverse resistance $R_{pp'}$ in the gap between (0↓-) and (1↑+) Landau levels in a n-channel Si(100) MOSFET with length to width ratio of 6. At gate voltages between 14.4 V and 14.6 V, $R_{pp'}$ passes through a minimum ($\simeq 5 \times 10^{-4}\Omega$) and R_H passes through a plateau valued R_H = 6453.2024 ± 0.0008 Ω. This result was not naken into account in the final result because this measurement was made without oil bath.[63]

effect or the quantum Hall effect[59∿61].

Several high precision measurements of the quantized Hall resistance of Si-MOSFETs have been reported[62∿66]. Results of high precision measurements of the quantized Hall resistance of GaAs-GaAlAs heterojunction have been also reported[67,68].

An example of Yamonouchi et al's[63] results of high precision measurements of a plateau value of Hall resistance R_H = 6453.2024 ± 0.0008 ohm and a minimum value of transverse resistance $R_{pp'} \simeq 5 \times 10^{-4}$ ohm in the gap between the (0↓-) and the (1↑+) Landau levels in a long Hall sample having a high electron mobility in 14 T at 1.4 K is shown in Fig. 16. Their final result of R_H and its one standard deviation uncertainty are

$$R_H = (6453.1998 \pm 0.0032)\Omega_{ETL} \quad (0.50 \text{ ppm})$$

or

$$R_H = (6453.1998 \pm 0.0041)\Omega_{ABS} \quad (0.64 \text{ ppm}).$$

This result has lead the inverse of the fine structure constant α^{-1} = 137.035956 (±0.64 ppm) which agrees with the value of α^{-1} so far obtained[69]. (See the Note Added at the end of this article.)

The accuracy of the measurements of the absolute value of the Hall resistance in Japan is mainly limited to the accuracy of the standard resistor in Electrotechnical Laboratory which is ±0.5 ppm. Such a systematic uncertainty can be eliminated when one measures the quantized Hall resistance $R_H(i) = h/ie^2$ for different integer values of i in the same sample by using the same experimental system and compares i x $R_H(i)$ with each other. Yoshihiro et al[70,71] have made high precision measurements of $R_H(i) = h/ie^2$ for i = 4, 8, 12 in the magnetic fields of 9.0 T and 10.5 T at temperatures around 0.5 K. Their results are summarized in Table 1. The ratios with one standard deviation uncertainties are given as

$$2 \times \bar{R}_H(8)/\bar{R}_H(4) = 0.99999993 \pm 0.00000018,$$
$$2 \times \bar{R}_H(8)_{9T}/\bar{R}_H(4) = 1.00000005 \pm 0.00000019,$$
$$3 \times \bar{R}_H(12)/\bar{R}_H(4) = 0.99999995 \pm 0.00000028.$$

Here $\bar{R}_H(8)_{9T}$ is the mean value of $R_H(8)$ measured in B = 9 T, and the uncertainties include the estimated systematic uncertainty of ± 0.15 ppm. Thus the quantized Hall resistance is found to be integral

TABLE 1. Summary of Hall resistance and transverse resistance data for the sample 72-17H53-17-L1. $\bar{R}_H(i)$: the average of $R_H(i)$ values obtained at V_g's where ρ_{xx} < 10 mΩ/sq; σ_{n-1}: one standard deviation of $R_H(i)$ data; ρ_{xx}^{min}: minimum value of ρ_{xx} (i). The electron concentration for the respective i and B is given by ≃ 2.4 x 10^{10} iB (in Tesla) cm^{-2}. (Ref. 70)

i	H (T)	$\bar{R}_H(i)$ (Ω)	σ_{n-1} (Ω)	n	Range of V_s of ρ_{xx}<10 mΩ(V)	ρ_{xx}^{min} (mΩ)
4	10.5	6453.2005	0.0011	11	10.46 -10.82	0.15
8	10.5	3226.6001	0.0007	6	21.24 -21.40	0.19
8	9.0	3226.6004	0.0006	3	18.20 -18.28	0.85
12	10.5	2151.0667	0.0009	3	31.98 -32.02	5.3

fraction of h/e^2 in precision better than 2 parts in 10^7 over the electron concentrations ranging from 1 x 10^{12} cm^{-2} (i = 4) to 3 x 10^{12} cm^{-2} (i = 12) for the case of high mobility Si-MOSFET samples.

Kinoshita et al[72] have made the measurements of the quantized Hall resistance and transverse resistance at temperatures down to 12mK in a magnetic field of 12 T. They have shown that the nimina of ρ_{xx}'s lower than 10^{-2} Ω are realized for i = 4, 6, 8 and 12 at the channel current I_{SD} = 10 μA and for i = 16 at I_{SD} = 2.5 μA. This fact shows that these Hall resistance plateaus are available as quantum standard resistors of five different values down to 1.6 KΩ (i = 16) on a single device in the field of 12 T at temperatures attained by use of a ^{3}He-^{4}He dilution refrigerator. They have also measured the range of the gate voltage where ρ_{xx} ≃ 0 which roughly corresponds to $N_{immobile}$ in FIG. 13. Their $N_{immobile}$ evaluated by the gate voltage region where σ_{xx} < 5 x 10^{-7} mho at 12 mK approximately fit to the empirical relations in FIG. 13; $N_{immobile}$ = (eB/h)/(2N+1).

Up to the present, no theoretical discussion has been made about the ultimate accuracy of the quantization of the Hall resistance in the experimentally available situations. Yoshihiro et al[73] have investigated experimentally limit of accuracy of the quantized Hall resistance in Si-MOSFETs determined from the characteristics of devices and the experimental conditions. They have measured the Hall and transverse resistances at temperatures ranging from 4.2 K to 12mK in the magnetic fields up to 12 T passing through the channel current ranging from 10^{-5} A to 10^{-9} A. They have found that (1) random peaked structures are superposed on both Hall and transverse resistance versus gate voltage curves, (2) the position in gate voltages of the peaks and dips is independent of temperature and magnitude of channel current, (3) the amplitude increases with decreasing the temperature and decreases with increasing the channel current, and the amplitude is roughly independent of the magnetic field. They have concluded that although the random structure is not appreciable under the condition of high precision measurements[70)], it will limit the accuracy of measurements when the resolution of the measurements is improved up to the order of 10^{-8}.

3-3 Temperature Dependence of the Localization in Landau Levels: Effective Mobility Edge and Localization Length

As is described in the former section, the localization in Landau levels in 2D systems plays very interesting roles in the Hall effect as shown in Fig. 15 and Fig. 16. At low temperatures, the electronic states for $\sigma_{xx} = 0$ produce the constant plateaus in the Hall conductivity against the change in electron concentration. The plateau values coincide with the integer multiple of e^2/h in high precision of less than 1 ppm. The Hall plateau has the electron-hole symmetry; i.e. the Hall conductivity is equal to $-Ne^2/h$ when the Fermi level lies in the localized states near the higher edge of the (N-1)th Landau level and when the Fermi level lies in the localized states near the lower edge of the Nth Landau level. We have measured temperature dependence of the Hall conductivity and the transverse conductivity, and we have analysed the experimental results on the basis of the Anderson localization picture.

Gate voltage dependences of the Hall conductivity σ_{xy} and the transverse conductivity σ_{xx} change with temperature as shown in Fig. 17[56)]. The behaviour of σ_{xy} and σ_{xx} in the lowest Landau level and in the gap region between the lowest two Landau levels is different from the behaviour in higher Landau levels and in higher gap regions. The electrons in the lowest Landau level look like to localized

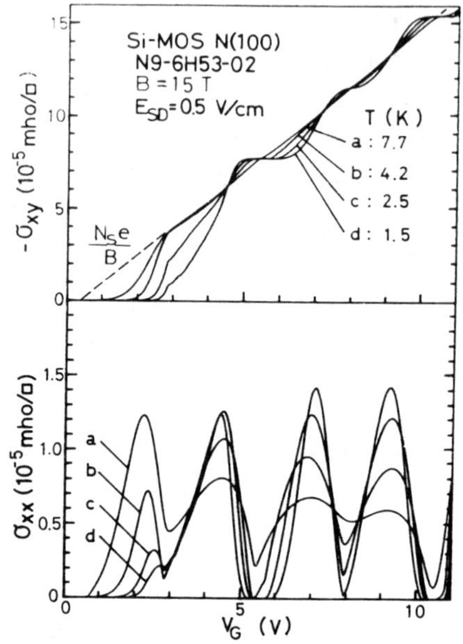

FIG. 17. Changes in Hall conductivity σ_{xy} and transverse conductivity σ_{xx} versus gate voltage V_G with temperatures in the N=0 Landau level in a Si MOSFET in Fig. 15.[56]

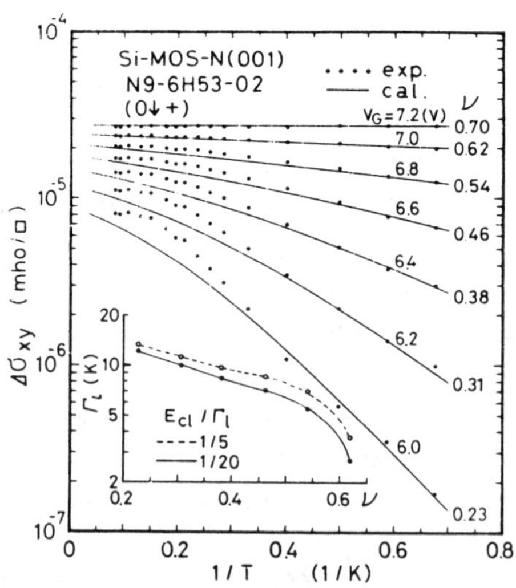

FIG. 18. Temperature dependence of $\Delta\sigma_{xy} = -\sigma_{xy}-(-2e^2/h)$ for gate voltages between 6.0V and 7.2V in the data in Fig. 17. ν is the occupation factor of the Ladau level $(0{\downarrow}+)$.[75]

completely at $T \to 0$ when the Fermi level lies in the lowest Landau level, i.e. $\sigma_{xy} = 0$ as well as $\sigma_{xx} = 0$ at absolute zero. However, at T = 1.5 K, a part of the electrons in the lowest Landau level becomes delocalized to carry Hall current corresponding to one Landau level, $-(e^2/h)E_y$, when the Fermi level rises up to fill the second Landau level more than half. This behaviour looks to show that the effect of screening is important in the localization. The nature of the localization in the lowest Landau level is probably different from that in higher Landau levels.

In the following, we will confine our analysis to the $(0\downarrow+)$ Landau level or the third lowest Landau level for the behaviour of this Landau level looks simplest among Landau levels in Fig. 17. We will confine further our analysis to the Hall conductivity only for the temperature dependence of σ_{xy} in this Landau level shows more clearly and simply the effect of the localization than that of σ_{xx}.

As is discussed by Aoki and Ando[74], and as shown by Ando's numerical experiments[50], the Hall conductivity in a Landau level increases almost linearly from zero to e^2/h against the Fermi level passing from the lower mobility edge to the higher mobility edge. Then, we can assume the change in the Hall conductivity in a Landau level as

$$\Delta\sigma_{xy}(\nu,\ T) = -\frac{e^2}{h}\cdot\frac{n_M(\nu,T)}{N_M} \tag{10}$$

where, ν is the filling factor or the occupancy factor of the Landau level, N_M the total number of the extended states and $n_M(\nu,\ T)$ the number of electrons in the extended states. Based on such a mobility edge model, Kawaji, Wakabayashi and Moriyama[75] have tried a phenomenological analysis of the temperature dependence of $\Delta\sigma_{xy}(T) = \sigma_{xy}(T) -(-2e^2/h)$ in the lower half of the $(0\downarrow+)$ Landau level in Fig. 17.

They have used the following model: The density of states in a Landau level is constant between the lower-edge $-\Gamma_\ell$ and the higher-edge Γ_h. The lower- and higher- mobility edge lie at $-E_{c\ell}$ and E_{ch}, respectively. All the electronic states are localized outside of these mobility edges. The ratio $E_{ch}/E_{c\ell}$ is assumed to be equal to Γ_h/Γ_ℓ. The ratio Γ_h/Γ_ℓ is determined by $\Gamma_h/\Gamma_\ell = 1/\nu_0 - 1$. Here, ν_0 is the filling factor of the Landau level at which σ_{xy} is independent of temperature. In the present case, $\nu_0 = 0.70$. The ratio $\Gamma_{c\ell}/\Gamma_{ch}$ is assumed to be independent of ν.

As shown in Fig. 18, experimental $\Delta\sigma_{xy}$ (dots) can be reproduced by eq(10) (solid lines) at temperatures lower than 2.5 K by adjusting Γ_ℓ for different gate voltages (different ν). The solid lines in Fig. 18 shows calculated $-\Delta\sigma_{xy}$ for $E_{c\ell}/\Gamma_\ell = 1/20$. However, experiments can be similarly reproduced by use of $E_{c\ell}/\Gamma_\ell < 1/5$. The inserted figure shows plots of Γ_ℓ as a function of ν for $E_{c\ell}/\Gamma_\ell = 1/5$ and 1/20.

The present simple mobility edge model can explain temperature dependence of σ_{xy} only at temperatures lower than 2.5 K. In order to describe the temperature dependence at higher temperatures, we have to introduce phenomenological mobility edges which move to the Landau level edges at higher temperatures; $E_{c\ell}/\Gamma_\ell \rightarrow 1$ at $T \gg 2.5$ K.

Recently, Ando[59] has extended the sample size in the numerical experiments and shown that there is no mobility edge in the N = 0 Landau level or the localization length diverges at the center of the Landau level. Ono[76] has obtained analitically similar results which show that the localization length is given by

$$\xi = 0.9437\ \ell \exp\left(\frac{\Gamma^2 + 4E_F{}^2}{8E_F{}^2}\right) \tag{11}$$

near the center of the Landau level.

We have to remind here that the effect of inelastic scattering plays an important role in the weak localization in the absence of strong magnetic fields. As is first discussed by Thouless, the sample size L should be replaced by the inelastic diffusion length $\sqrt{D\tau_\varepsilon}$ when $\sqrt{D\tau_\varepsilon} < L$ which is usually satisfied. Therefore, the electrons can not be localized when the localization length ξ is longer than the inealstic diffusion length $\sqrt{D\tau_\varepsilon}$. Thus an effective mobility edge exists at E_c where $\xi(E_c) = \sqrt{D\tau_\varepsilon}$. Therefore, it is expected that the temperature dependence of the inelastic scattering time τ_ε controls the temperature dependence of the effective mobility edge.

Moriyama and Kawaji[77] have discussed temperature dependence of the effective mobility edge derived from the temperature dependence of $d\sigma_{xy}/dN_s$ or $d\sigma_{xy}/dV_G$.

In the N = 0 Landau levels, the electron localization is so strong that the contribution of σ_{xx} term to σ_{xy} is very small in high mobility samples. Therefore, $d\sigma_{xy}/dN_s$ is a useful quantity in analysis of experiments on electron localization in Landau levels in 2D systems. When there is no localization, $d\sigma_{xy}/dN_s = -e/B$ in the case where the σ_{xx} term in σ_{xy} can be ignored. When the localization is strong $d\sigma_{xy}/dN_s \simeq 0$ at low temperatures in the region where the Fermi level lies in the localized states.

Figure 19 shows the gate voltage dependence of $-d\sigma_{xy}/dN_s$ as well as σ_{xx} and σ_{xy} in four N = 0 Ladau levels at T = 1.44 K in B = 15 T. Temperature dependence of $d\sigma_{xy}/dN_s$ at two different gate voltages where the filling factors of the $(0\downarrow+)$ Landau level are $\nu = 0.187$ and 0.555 are shown in Fig. 20 .

In order to obtain the effective mobility edges, Moriyama and Kawaji have assumed a following function for the density of states in a Landau level:

$$D(E) = \frac{1}{2\pi\ell^2} \cdot \frac{3\cdot 2^{2/3}/\Gamma(1/3)}{\Gamma_\ell+\Gamma_h} \; e^{-(E/\Gamma_i)^3} \tag{12}$$

where Γ_ℓ and Γ_h are the half width of the Landau level in the lower and the higher side, respectively, and $\Gamma_i = \Gamma_\ell$ for $E < 0$ and $\Gamma_i = \Gamma_h$ for $E > 0$. Two mobility edges, the lower mobility edge and the higher mobility edge, lie at $E = -E_{c\ell}$ and $E = E_{ch}$, respectively. The ratio $E_{ch}/E_{c\ell}$ is assumed to be equal to Γ_h/Γ_ℓ. The ratio Γ_h/Γ_ℓ is determined by $\Gamma_h/\Gamma_\ell = 1/\nu_0 - 1$ where ν_0 is the filling factor of the Landau level at which σ_{xy} is independent of temperature. Then we can determine $E_{c\ell}/\Gamma_\ell$ which reproduces the derivative $d\sigma_{xy}(\nu,T)/dN_s$ obtained in experiments at each temperature.

Figure 21 shows temperature dependence of the effective mobility edge E_c/Γ obtained from the temperature dependence of $d\sigma_{xy}/dN_s$ at two different gate voltages. In this analysis, the level broadening in the self-consistent Born approximation by Ando and Uemura[54] for electron mobility of 10,800 $cm^2/v\cdot s$, $\Gamma = 20$ K, is employed for Γ_ℓ. The Fermi level lies at $E_F/\Gamma_\ell \simeq -0.5$ for $\nu = 0.187$ and $E_F/\Gamma \simeq 0$ for $\nu = 0.555$. Both curves in Fig. 21 appear to approach to $E_c/\Gamma = 0$ at $T \to 0$ as has been expected from numerical results by Ando[59] and analytical results by Ono[76].

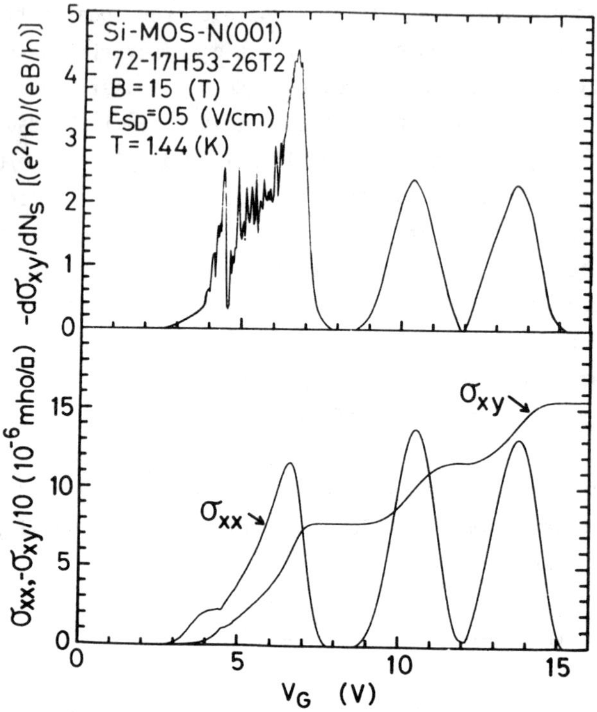

FIG. 19. Hall conductivity σ_{xy}, transverse conductivity σ_{xx} and the derivative of σ_{xy} with electron concentration N_S, $-d\sigma_{xy}/dN_S$, in the N=0 Landau level.[77]

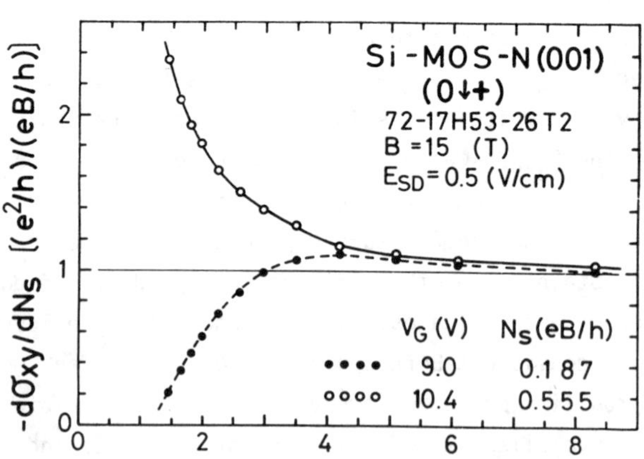

FIG. 20. Temperature dependence of $-d\sigma_{xy}/dN_S$ at two different occupation factors in the (0↓+) Landau level in Fig. 19.[77]

We can evaluate now the localization length ξ at E_c/Γ by employing Ando's results or Ono's results. Experimentally obtained effective mobility edge in Fig. 21 is far from E = 0. Therefore, Ando's numerical results is better than Ono's analytical results to derive the localization length from the experimental results of E_c/Γ. Figure 22 shows the localization length $\xi(E_c/\Gamma)$ as a function of 1/T derived from our experimental results in Fig. 19 and Ando's results of numerical experiments (Fig. 12 in ref. 59). The magnitude and the temperature dependence of the localization length in Fig. 21 appear not unreasonable.

If the effective mobility edge $E_c(T)/\Gamma$ is determined by $\sqrt{D\tau_\varepsilon}(T) = \xi(E_c/\Gamma)$ disscused above, we can evaluate $\sqrt{D\tau_\varepsilon}$ as a function of temperature. The inelastic diffusion length is well approximated by $L_\varepsilon = L_{\varepsilon 0}e^{\Delta/T}$. The diffusion constant in a Landau level in two-dimensional systems is approximately given by $D \simeq \ell^2/\tau$ where ℓ is the radius of the Landau orbit and τ the elastic scattering time. Thus the results in Fig. 22 give a result

$$\frac{\tau_\varepsilon}{\tau} \simeq e^{T_\varepsilon/T} \quad . \tag{13}$$

where $T_\varepsilon \simeq 7.1$ K at $\nu = 0.555$ and $T_\varepsilon = 6.4$ K at $\nu = 0.187$. This is a temperature dependence of the inelastic scattering time which have not observed in the negative magnetoresistance experiments in the weakly localized regime.

Let us consider another possibility of the effective mobility edge. If the inelastic scattering is short, the level broadening determined by $\delta E \simeq h/\tau_\varepsilon$ controls the effective mobility edge E_c as $E_c = \hbar/\tau_\varepsilon$. Results in Fig. 20 gives $\tau_\varepsilon/\tau = 40 \sim 1000$ which leads $\delta E/\Gamma \ll E_c/\Gamma$. Therefore, the possibility of $E_c/\Gamma \simeq \tau/\tau_\varepsilon$ can be ignored.

In conclusion the temperature dependence of the effective mobility edge obtained by analysis of the Hall conductivity measurements in Si MOSFETs at temperatures down to 1.5 K in a magnetic field of 15 T based on a simple model expressed by eq(10) appears to lead us to a conclusion that there is no mobility edge in 2D systems in strong magnetic fields. This conclusion agrees with numerical results of Ando[59)] and analytical results of Ono[76)]. Although we have no

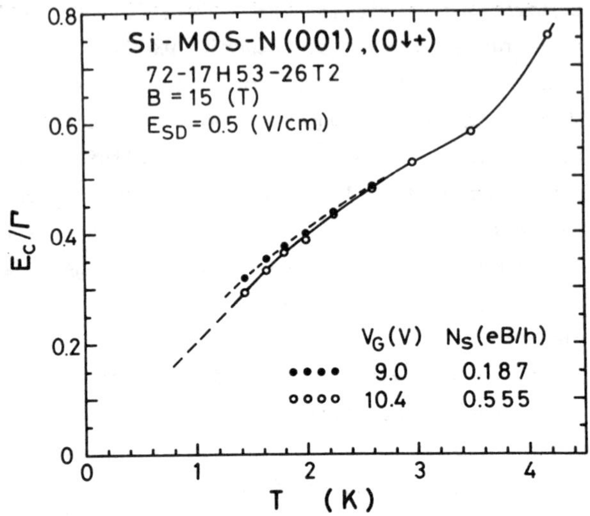

FIG. 21. Temperature dependence of the effective mobility edge $E_c/\Gamma = E_{c\ell}/\Gamma_\ell = E_{ch}/\Gamma_h$ deduced from the data in Fig. 20.[77)]

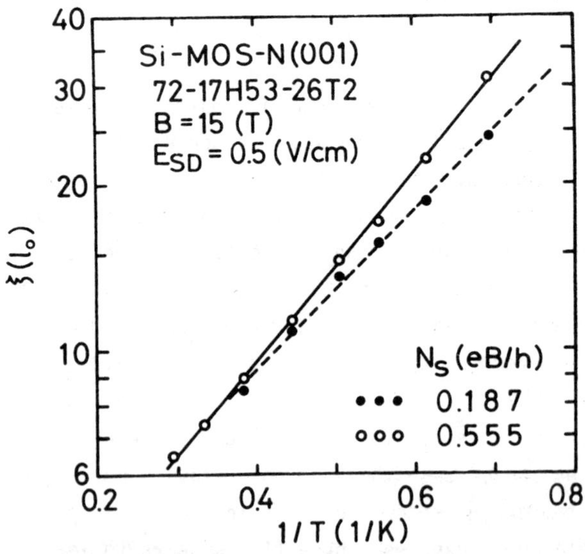

FIG. 22. Temperature dependence of the localization length ξ in unit of the radius of ground Landau orbit ℓ_0 deduced from E_c/Γ in Fig. 21 by use of Ando's results of $\xi(E/\Gamma)$.

theoretical explanation yet for the inelastic diffusion length shown in Fig. 21 or the inelastic scattering time given by eq(13) which is derived by use of Ando's reslults for the localization length in the N = 0 Landau level, the inelastic diffusion length is expected to exceed the sample size of 100 μm at a certain temperature higher than 0.1 K. Further experimental studies at lower temperatures than 1 K are remained to be done.

4. Summary and Conclusion

We have reviewed experimental researches on electron localization in silicon MOS inversion layers in weak and strong magnetic fields performed in our group. Results of negative magnetoresistance experiments in the weakly localized regime have been well explained by recently developed theories based on Anderson localization in two-dimensional systems and effect of mutual Coulomb interactions on the orbital motion in two-dimensional systems including effects of inter-valley impurity scattering. Based on hot-electron effect experiments on negative magnetoresistance we have reached an important conclusion that the deformation potential constants in electron-phonon interaction in inversion layers in the Si-SiO_2 interface are the same to those in bulk silicon.

Most interesting behaviour of two-dimensional systems in random potentials has appeared in strong magnetic fields where the Hall conductivity is quantized to an integer multiple of $-e^2/h$ in precision better than 2 parts in 10^7. We have confirmed the electron-hole symmetry in the localization in Landau levels which produce the same integer multiple of $-e^2/h$. Our simple analysis of the Hall conductivity has appeared to show that no mobility edge exists in a Landau level in two-dimensional systems though our experiments have been done at temperatures higher than 1.5 K. Electronic states near the center of a Landau level appear to extend at low temperatures. However, further experiments at lower temperatures should be done before arriving at the final conclusion.

It is interesting to note here that both phenomena — the negative magnetoresistance in weak magnetic fields and the quantization of Hall resistance in strong magnetic fiels — are based on the quantum mechanical nature of the wave functions and energy levels of electrons which depend on the vector potential of the magnetic fields (Aharanov-Bohm effect).

In this review we have not discussed on researches on strongly localized regime in silicon MOS inversion layers. Though we have done some attempts to understand the properties in this regime in the light of the scaling arguments[78,79], we have no theory at finite temperatures in this regime. We hope to find solution to this problem on a line extended from the researches which have been performed on weak magnetic field effects in weakly localized regime and on the strong localization in Landau levels.

Acknowledgements

This review owes to the authors' colleagues whose works are presented in many figures. The author is grateful to Dr. A. Yagi, Sony Corporation, who supplied them with the greater part of Si MOSFETs and Dr. T. Sato, Toshiba Research and Development Center, who supplied them with Si MOSFETs used in the work presented in Fig. 14. The author is also grateful to Professor S. Tanuma, Dr. H. Suematsu and Mr. I. Oguro who assisted them in the use of high magnetic field facilities in the Institute for Solid State Physics, University of Tokyo. The author is indebted to Drs. H. Fukuyama, T. Ando, Y. Ono, A. Kawabata and K. Nakamura for discussions during preparation of this article. In the last but not the least, the author would like to thank Professor Y. Uemura for his encouragement through the work reviewed in this article. A part of this work has been supported by the Grants-in-Aid for the Scientific Research from the Ministry of Education.

References

1) T. Ando, A. B. Fowler, and F. Stern, Rev. Mod. Phys. 54,437(1982).
2) Y. Uemura, Proc. 2nd Int. Conf. on Solid Surfaces, Kyoto, 1974; Japan. J. Appl. Phys. Suppl. 2, Pt. 2, 17 (1974).
3) S. Kawaji and Y. Kawaguchi, Proc. 9th Int. Conf. Phys. Semiconductors, Moscow, (Nauka, Leningrad), p. 730, 1968.
4) S. Tansal, A. B. Fowler and A. B. Cotellessa, Phys. Rev. 178,1326 (1969).
5) H. Sakaki and T. Sugano, J. Japan Soc. Appl. Phys. 41 suppl., 141

(1972).

6) I. Eisel and G. Dorda, Phys. Rev. Lett. 32, 1360 (1974).

7) S. Pollitt, M. Pepper, C. J. Adkins, Surface Sci. 58, 79 (1976).

8) Y. Kawaguchi, H. Kitahara and S. Kawaji, Solid State Commun. 26, 701 (1978).

9) Y. Kawaguchi, H. Kitahara and S. Kawaji, Surf. Sci. 73,520(1978).

10) Y. Goldstein, Y. Grishpun and A. Many, Phys. Rev. B19,2256(1979).

11) Y. Uemura and Y. Matsumoto, J. Japan Soc. Appl. Phys. 40 Suppl, 205 (1970).

12) E. Abrahams, P. W. Anderson, D. C. Licciardello and T. V. Ramakrishnan, Phys. Rev. Letters 42, 673 (1979).

13) P. W. Anderson, E. Abrahams and T. V. Ramakrishnan, Phys. Rev. Letters 43, 718 (1979).

14) D. J. Thouless, Phys. Rev. Letters 39, 1167 (1977).

15) G. J. Dolan and D. D. Osheroff, Phys. Rev. Letters 43, 721 (1979).

16) D. J. Bishop, D. C. Tsui, and R. C. Dynes, Phys. Rev. Letters 44, 1153 (1980).

17) S. Hikami, A. I. Larkin and Y. Nagaoka, Prog. Theoret. Phys. 63, 707 (1980).

18) B. L. Altshuler, D. E. Kmel'nizkii, A. I. Larkin and P. A. Lee, Phys. Rev. B22, 5142 (1980)

19) Y. Kawaguchi and S. Kawaji, J. Phys. Soc. Jpn. 48, 699 (1980).

20) Y. Kawaguchi and S. Kawaji, J. Phys. Soc. Jpn. 49 Suppl. A, 983 (1980).

21) Y. Kawaguchi and S. Kawaji, Surf. Sci. 113, 505 (1982)

22) M. J. Uren, R. A. Davis, M. Kaveh and M. Pepper, J. Phys. C. 14, L 395 (1981).

23) R. G. Wheeler, Phys. Rev. B24, 4645 (1981).

24) D. J. Bishop, R. C. Dynes, and D. C. Tsui, Phys. Rev. 26,773(1982).

25) S. Maekawa and H. Fukuyama, J. Phys. Soc. Jpn. 50, 2516 (1980).

26) H. Fukuyama, Prog. Theor. Phys. Suppl. 69, 220 (1981).

27) B. L. Altshuler, A. G. Aronov and P. A. Lee, Phys. Rev. Letters 44, 288 (1980).

28) H. Fukuyama, J. Phys. Soc. Jpn. 48, 2169 (1980).

29) M. Kaveh, J. J. Uren, R. A. Davies and M. Pepper, J. Phys. C14, L 413 (1981).

30) M. Kaveh and N. F. Mott, J. Phys. C14, L 177 (1981).

31) T. Tsuzuki, Solid State Commun. 38, 915 (1981).
32) T. Tsuzuki, Solid State Commun. 40, 339 (1981).
33) B. L. Altshuler, A. G. Aronov, D. E. Kmel'nizkii, Solid State Commun. 39, 619 (1981).
34) T. Tsuzuki, Solid State Commun. 44, 1175 (1982).
35) H. Fukuyam, Surface Sci. 113, 489 (1982).
36) H. Fukuyama, "Anderson Localization" ed. by Y. Nagaoka and H. Fukuyama, (Springer-Verlag, Berlin, 1982) p. 89.
37) A. A. Abrikosov, "Lecture on The Theory of Normal Metals" (Hidustan Publishing, India, 1968) p. 24.
38) E. Abrahams, P. W. Anderson, P. A. Lee, and I. V. Ramakrishnan, Phys. Rev. B24, 6783 (1981).
39) Y. Shinba, K. Nakamura, M. Fukuchi, and M. Sakata, J. Phys. Soc. Jpn. 51, 157 (1982).
40) T. Ando, Surf. Sci. 98, 327 (1980).
41) W. L. Bloss, L. J. Sham, and B. Vinter, Surface Sci. 98, 250(1980).
42) Y. Kawaguchi and S. Kawaji, Proc. 16th Int. Conf. on Phys. of Semicond., Montpellier, 1982, to be published.
43) Y. Kawaguchi and S. Kawaji, Japan. J. Appl. Phys. 21, L 709 (1982).
44) Y. Shinba and K. Nakamura, J. Phys. Soc. Jpn. 50, 114 (1981).
45) Y. Matsumoto and Y. Uemura, Proc. 2nd Int. Conf. on Solid Surfaces, Kyoto, 1974; Japan. J. Appl. Phys. Suppl. 2, Pt. 2, 367 (1974).
46) S. Kawaji, Surface Sci., 73 46 (1978).
47) A. B. Fowler, F. F. Fang, W. E. Howard, and P. J. Stiles, Phys. Rev. Letters 16, 901 (1966).
48) S. Kawaji and J. Wakabayashi, Surface Sci. 58, 238 (1976).
49) T. Ando, Surface Sci. 113, 182 (1982).
50) T. Ando, "Anderson Localization" ed. by Y. Nagaoka and H. Fukuyama (Springer-Verlag, Berlin, 1982), p. 176.
51) T. Ando, Y. Matsumoto and Y. Uemura, J. Phys. Soc. Jpn. 39, 279 (1975).
52) T. Igarashi, J. Wakabayashi, and S. Kawaji, J. Phys. Soc. Jpn. 38, 1549 (1975).
53) S. Kawaji, T. Igarashi and J. Wakabayashi, Prog. Theor. Phys. Suppl. 57, 176 (1975).
54) T. Ando and Y. Uemura, J. Phys. Soc. Jpn. 36, 959 (1974).
55) J. Wakabayashi and S. Kawaji, J. Phys. Soc. Jpn. 48, 333 (1980).

and Surface Sci. 98, 299 (1980).
56) S. Kawaji and J. Wakabayashi, "Physics in High Magnetic Fields" ed. by S. Chikazumi and N. Miura (Springer, Berlin, 1981) p.284.
57) T. Ando, private communication (1974).
58) K. von Klitzing, G. Dorda, and M. Pepper, Phys. Rev. Letters 45, 494 (1980).
59) T. Ando's article in this volume and references cited therein.
60) M. E. Cage and S. M. Girvin, Comments on Solid State Physics, to be published.
61) S. M. Girvin and M. E. Cage, Comments on Solid State Physics, to be published.
62) E. Braun, E. Staben, and K. von Klitzing, PTB Mitl. 90, 350(1980).
63) C. Yamanouchi, K. Yoshihiro, J. Kinoshita, K. Inagaki, J.Moriyama, S. Baba, S. Kawaji, K. Murakami, T. Igarashi, T. Endo, M.Koyanagi and N. Nakamura, "Precision Measurements and Fundamental Constants II ", ed. by B. N. Taylor and W. D. Phillips (Nat. Bur. Stand. U. S., Spec. Publ. 617, to be published).
64) M. E. Cage, R. F. Dziuba, B. F. Fields, C. F. Lavine, and R. J. Wagner, ibid., to be published.
65) A. Hartland, ibid., to be published.
66) K. Yoshihiro, J. Kinoshita, K. Inagaki, C. Yamanouchi, J.Moriyama, and S. Kawaji, Surface Sci. 113, 16 (1982).
67) R. F. Dziuba, M. E. Cage, B. F. Field, D. C. Tsui, and A. C. Gossard, "Precision Measurements and Fundamental Constants II ", ed. by B. N. Taylor and W. D. Phillips (Nat. Bur. Stand. U. S. , Spec. Publ. 617, to be published).
68) D. C. Tsui, A. C. Gossard, B. F. Field, M. E. Cage and R. F. Dziuba, Phys. Rev. Letters 48, 3 (1982).
69) E. R. Williams and P. T. Olsen, Phys. Rev. Letters 42, 1575 (1979).
70) K. Yoshihiro, J. Kinoshita, K. Inagaki, C. Yamanouchi, J.Moriyama, and S. Kawaji, J. Phys. Soc. Jpn. 51, 5 (1982).
71) K. Yoshihiro, J. Kinoshita, K. Inagaki, C. Yamanouchi, J. Moriyama and S. Kawaji, Proc. 16th Int. Conf. on Semicond., Montpellier, France, 6-10 September, 1982 (to be published).
72) J. Kinoshita, K. Inagaki, C. Yamanouchi, K. Yoshihiro, J. Moriyama and S. Kawaji, Proc. Int. Conf. on Application of High Magnetic

Fields in Semiconductors Physics, Grenoble, Spetember 13-17, 1982 (to be published).

73) K. Yoshihiro, J. Kinoshita, K. Inagaki, C. Yamanouchi, K. Shida, T. Igarashi, J. Moriyama and S. Kawaji, Proc. 1982 Int. Conf. on Solid State Devices, Tokyo, August (to be published).

74) H. Aoki and T. Ando, Solid State Commun. 38, 1079 (1981).

75) S. Kawaji, J. Wakabayashi, and J. Moriyama, J. Phys. Soc. Jpn. 50, 3839 (1981).

76) Y. Ono, J. Phys. Soc. Jpn. 51, 2055 (1982).

77) J. Moriyama and S. Kawaji, Solid State Commun. 45 (1983) 511.

78) S. Kawaji, M. Namiki and N. Hoshi, J. Phys. Soc. Jpn. 49, 1637 (1980).

79) N. Hoshi and S. Kawaji, Surf. Sci. 113, 189 (1982).

80) J. Kinoshita, K. Inagaki, C. Yamanouchi, K. Yoshihiro, J.Moriyama and S. Kawaji, Report submitted to 16^{e} Session, Comite Consultatif d'Electricite (March, 1983).

Note Added[80)]

The $R_H(4)$ value reported in Refs. 63 and 66 has been corrected as the value cited in this article based on the recent calibration of 10 KΩ resistor. The average 6453.1969 Ω should read 6453.1998 Ω (add 0.45 ppm).

Recent result gives the value and 1σ uncertainty as $h/4e^2$=6453.2008 ± 0.0021 $\Omega_{ETL,Nov.1982}$ (0.32 ppm); the corresponding fine structure constant is $1/\alpha$ = 137.035976 ± 0.000070 (0.51 ppm). The reproducibility of the $h/4e^2$ values is better than 0.1 ppm for 42 days. The result suggests that there should be a drift, which is about -0.06μΩ a year, in the magnitude of the as maintained unit Ω_{ETL}.

ELECTRONIC PROPERTIES OF GRAPHITE INTERCALATION COMPOUNDS

Sei-ichi Tanuma

Faculty of Engineering
Gunma University
Kiryu, Gunma, 376 Japan

ABSTRACT

Graphite is a layered crystal of which intrinsic carriers are as few as $4\times10^{18}cm^{-3}$ and yields various kinds of intercalation compounds. These compounds are synthesized as metals due to the charge transfer between the host and intercalants. This article reviews recent development of such electronic properties as resistivity, magnetism, superconductivity, optical spectra and magnetoplasma wave response. Problems on determining the electronic structure are mentioned.

1. Introduction

Graphite is one of the elemental semimetals in which the charge cariers are compensated, having the concentration of 4×10^{18} cm^{-3}. Due to the nature of layered structure of this material, many kinds of atomic and molecular species can intercalate into interlayer spaces, combine with graphite planes by the charge transfer and make intercation componds. Graphite intercalation compounds (GIC's) are thus kinds of synthetic metals if the intercalant was chosen in the view point of the character of species and the concentration. The most distinguished character of GIC is the staging; the n-th stage compound has such a super-structure that every intercalant layer is separated by n-layers of graphite. The systematic change of electronic properties of GIC's from the pristine graphite ones due to the stage advancement is quite drastic, while some of the basic characters of the host crystal is not lost. The change of electronic properties by intercalation is more or less similar to the change by impurity doping in a semiconductor, but the superlattice structure of intercalants by the staging maintains the high mobility of transferred electrons or holes in graphite π - band, giving superior values of electrical conductivity along the planary direction. From the view point of a new synthetic material, GIC shows quite interesting properties of structural, chemical, electrical, magnetic, optical and thermal, and makes a microcosmos of solid state science.

In this article, some of the novel electronic properties of GIC with which the author and/or his colleagues have been concerned are described, while a general feature is briefly mentioned. The items which are described are followings:

§2 electronic structure and transport of pristine graphite,
§3 electric conduction and magnetoresistance of GIC's,
§4 magnetism of intercalants and of conduction carriers,
§5 superconductivity of alkali metal- and alkali metal amalgam- GIC's,
§6 optical spectra and magnetoplasma wave, and
§7 problems on electronic structure of GIC's.

2. Electronic Structure and Transport of Pristine Graphite

2-1 Electronic Structure

Graphite crystal has a remarkable anisotropy between the in-plane C - C distance, 1.421 Å and the inter- planer distance, 3.354 Å, thus the first approximation is the single honeycomb carbon network of two dimension. Fig.1 shows the two dimensional energy band[1)]; σ - bands are antibonding- and bonding- sp^2 hybridized bands, and π - bands are corresponding to the rest of sp^3 carbon valece electrons.The Fermi level meets the P - point where the two empty and filled π - bands degenerate, so that the material should be a zero - gap semiconductor.

The actual graphite consists of ABAB··· stacking sequence of honeycomb planes as shown in the insert of Fig.20, and the unit cell contains four atoms instead of two in the two dimensional approximation. The detailed energy structure was obtained by the formuration by Slonczewski, Weiss [2)] and McClure[3)], which is called SWMcC - model. The Hamiltonian is as follows:

$$H = \begin{pmatrix} E_1 & 0 & H_{13} & H_{13}^* \\ 0 & E_2 & H_{23} & -H_{23}^* \\ H_{13}^* & H_{23}^* & E_3 & H_{33} \\ H_{13} & -H_{23} & H_{33}^* & E_3 \end{pmatrix}, \qquad (1)$$

where the band edge energies E_1, E_2, E_3 and the interaction terms H_{ij} are the followings:

$$E_1 = \Delta + \gamma_1 S + 1/2\ \gamma_5 S^2 ,$$

(a)

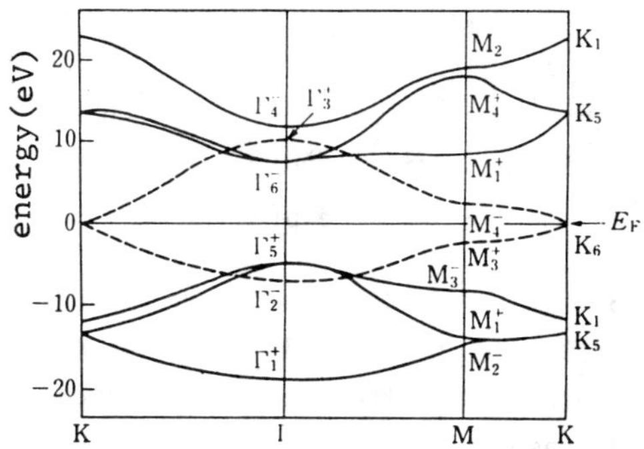

(b)

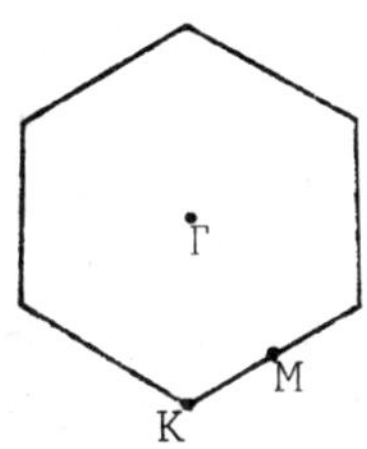

Fig. 1

(a) Band structure of the two dimensional graphite. (Tsukada et al.[1])
The full lines show σ-bands and the dashed lines show π-bands.

(b) Brillouin zone of the two dimensional graphite.

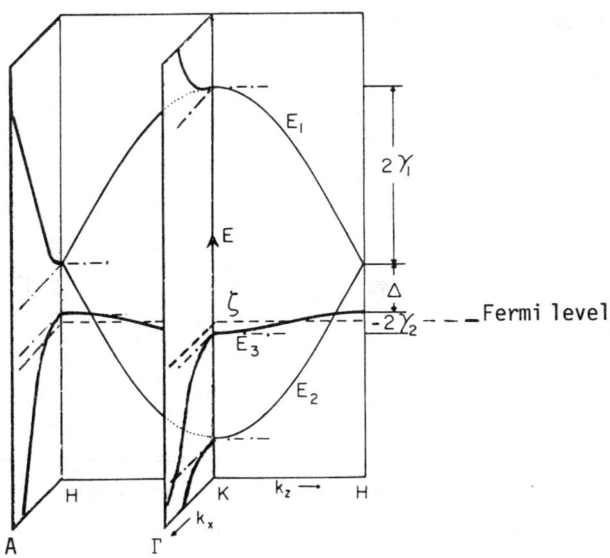

Fig.2 Schematic view of the band structure of three dimensional graphite, after McClure.[3] The minimum of E_3 band is located at the point K, because that the sign of γ_2 is decided as minus.

$$E_2 = \Delta - \gamma_1 S + 1/2\,\gamma_5 S^2, \qquad (2)$$
$$E_3 = 1/2\,\gamma_2 S^2,$$

$$\left.\begin{aligned} H_{13} &= 2^{-1/2}(-\gamma_o + \gamma_4 S)\,\eta\exp(i\theta), \\ H_{23} &= 2^{-1/2}(\gamma_o + \gamma_4 S)\,\eta\exp(i\theta), \\ H_{33} &= \gamma_3 S\,\eta\exp(i\theta) \end{aligned}\right\} \qquad (3)$$

and $S = 2\cos\pi\xi$, $\xi = k_z c_o/\pi$, $\eta = 1/2\cdot\sqrt{3}\,a_o/\kappa$. The angle θ is the one between k and ΓK directions, and the inter-atomic spacings c_o and a_o are 3.354 Å and 2.461 Å, respectively.

The parameters γ_o, γ_1, γ_2, γ_3, γ_4, γ_5 and Δ are the overlapping and transfer integrals between the nearest neighbor atoms, and the second nearest neighbor atoms, and the values of them have been determined by the comparizon with experiments. On the other hand, numerous detailed calculations of the energy structure without parametrization have been published; among them the method developed by Uemura's group[4,5] is regarded as the most precise one. The numerical values of the band parameters calculated by them are listed in Table 1 in comparizon with the experimentally determined values.

TABLE 1. Theoretical and experimental values of band parameters in SWMcC model.

	theoretical 5)	experimental
γ_o	2.37 eV	2.7 ~ 3.2 eV
γ_1	0.32	0.3 ~ 0.4
γ_2	-0.019	-0.018 ~ -0.023
γ_3	0.29	0.2 ~ 0.3
γ_4	0.15	0.12 ~ 0.25
γ_5	0.021	$\lvert\gamma_5\rvert < 0.02$
Δ	-0.017	-0.02 ~ 0.02

Fig.2 shows a schematic three dimensional graphite band near the Brillouin zone edge HKH. As the number of atoms in the unit cell is doubled, the number of π-bands is four in this case, and the degeneracy at the point K in Fig.1 is now broken, instead, a small band overlap arises along HKH-line to make the material semimetallic. The Fermi surface and its location in the Brillouin zone are illustrated in Fig.3. The real Fermi surface is much thinner than the illustration, so that

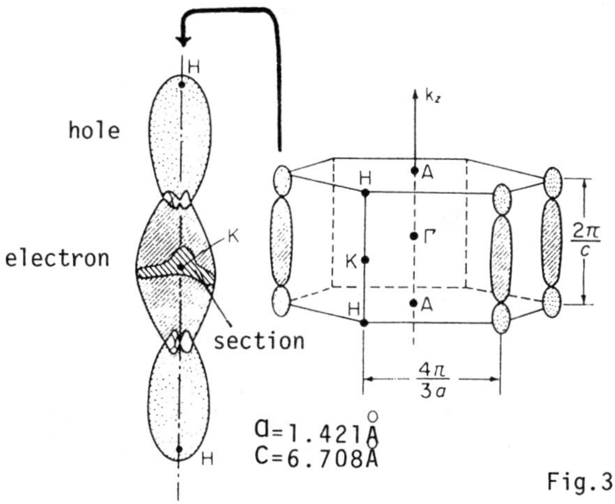

Fig.3

Brillouin zone and Fermi surface of the three dimensional graphite.
The diameter of Fermi surface compared to the length is much thinner than this illustration.

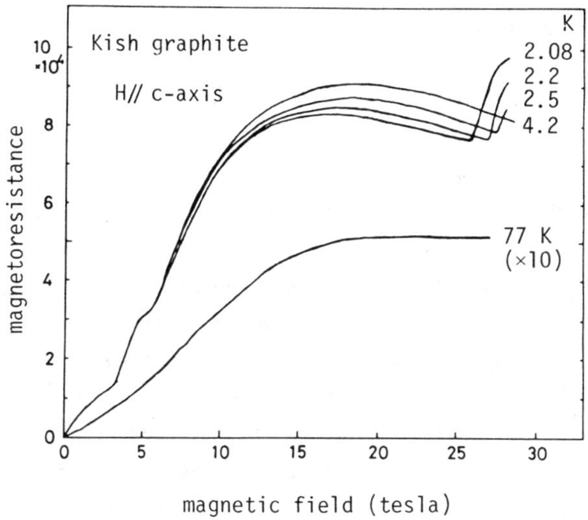

Fig.4
Anomalous increase of transverse magnetoresistance of single crystal graphite at lower temperatures than than 2.5 K and in higher magnetic fields than 25 T, by the use of pulsive magnetic field.
(Tanuma et al.[8,9])

the motion of the carriers are pseudo-two dimensional.

2-2 Magnetoresistance

The transverse magnetoresistance (MR) of a semimetal is very large, because of the lack of Hall field. Pure graphite shows a large MR; the value of (ρ_H - ρ_o)/ ρ_o amounts to 10^4 - 10^5 at 10 T at the liquid helium temperature and tends to saturate at higher magnetic fields. A well known peculiarity is the linear relationship between the MR and the field strength, instead of the ordinary H-squared relation, near or in the quantum limit region. This behavior was explained by McClure et al.[6] in a following way.The transverse magnetoconductivity is proportional to the number of scattering center of ionized impurity n_s, the reciprocal of the cyclotron frequency ω and the integration $\int dE \cdot (df/dE)[N(E)]^2 \langle V^2 \rangle$, where N(E) isthe state density and $\langle V^2 \rangle$ is the cross section of impurity scattering. The proportionality, $V \propto [N(E)]^{-1}$, is proved after considering the special character of Landau level in the quantum limit region as well as the screening of scattering, so that the remaining field dependent factor is ω ($\propto$ H) alone.

As for the trend to saturate (or even decrease for the purest specimens) with the field increase, Sugihara and Woolam[7] mentioned the effect of freezing-out of impurity levels; a conversion of ionized impurity to neutral one due to the magnetic field makes the force range shorter.

Such a behavior as mentioned above is shown in Fig.4. Contrary to an imagination that no more different change may occur in higher fields, a sudden increase of MR was found at and above 25 T at the temperature lower than 2.1 K, as shown as the right side of the MR curves in Fig.4. The original data with the use of pulsive magnetic field obtained by the authors[8,9] is copied in Fig.5. In order to prove that the observed anomary is intrinsic and to get more precise data, the MR was measured by Iye et al. using a 29Tstatic hybrid magnet in National Magnet Laboratory in M.I.T.[10] Thus the critical boundary of H_c vs. T for the occurrence of the anomarous increase of MR was obtained as shown in Fig.6. The mismatching between pulsive field data and stationary field data is thought as an eddy current effect in the sample in the pulse experiment, i.e., the real temperature is ca.0.5 K higher than the bath temperature which is plotted in the abscissa. The mechanism of this anomaly has

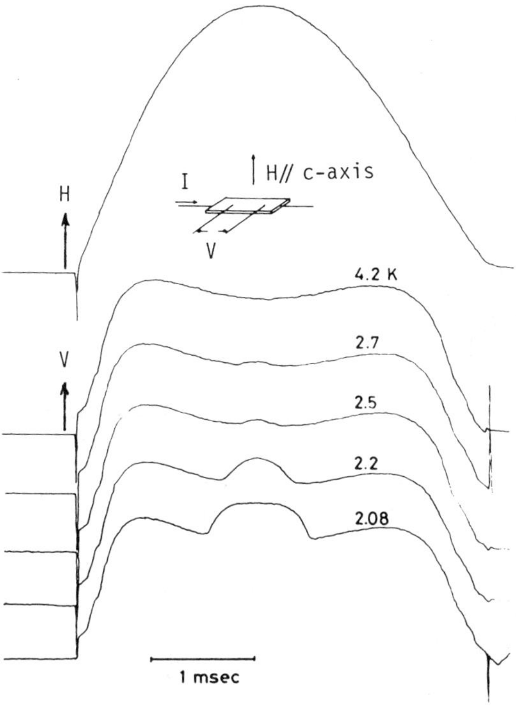

Fig.5
Copy of the original pulsive measurement of magnetoresistance. The time-dependences of magnetic field and voltage drop due to magnetoresistance are taken in parallel. Sudden increase of V is seen in the lower four curves.

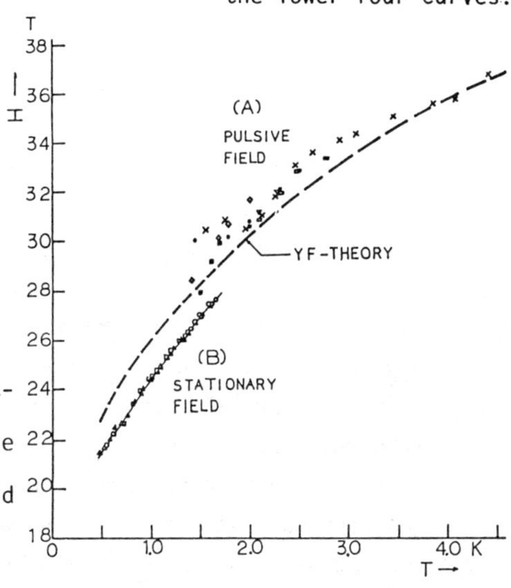

Fig.6
Plot of the threshold field for sudden increase of resistance versus the temperature. (a) is the plot from the pulse measurement and (B) is the plot from the stationary field measrement. The dashed line is the fit of Yoshioka-Fukuyama theory of CDW.

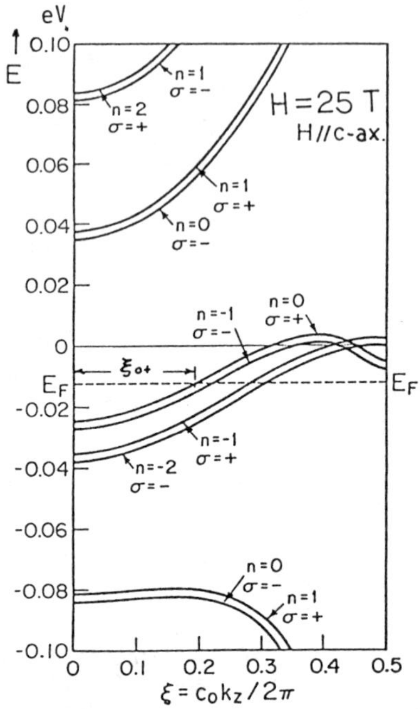

Fig.7
The Landau levels of graphite under the magnetic field parallel to the c-axis, of the intensity of 25 T. (Y.Iye et al.[10])
The Fermi level intersects only four levels, thus these four crossing points correspond to the Fermi surface, in a pseudo-one-dimensional way.

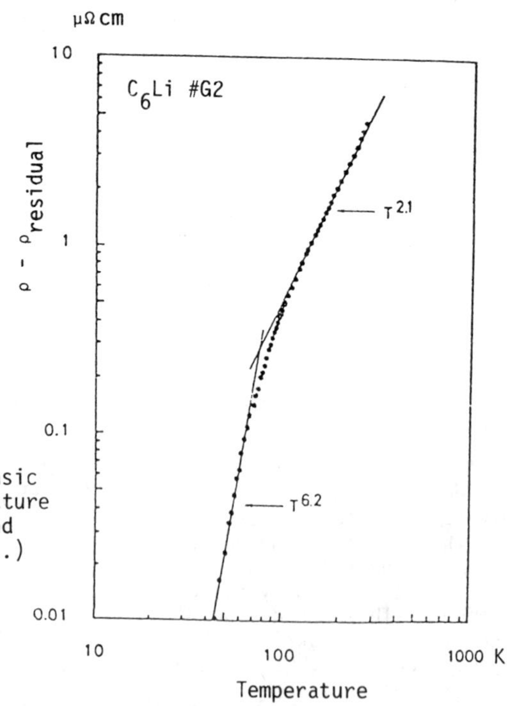

Fig.8
The log-log plot of intrinsic resistivity versus temperature on the first stage compound C_6Li. (after Y.Koike ey al.)

been proposed by Yoshioka and Fukuyama[11)] as the magnetic-field-induced charge density wave (CDW). This YF-theory is briefly introduced as follows. The Landau level at 25 T is calculated by the SWMcC-model giving the result shown in Fig.7. Only four levels concern the Fermi level, so that the electronic motion within the layer plane becomes localized and the conduction system becomes very much one dimensional. In YF-theory, various types of electronic instabilities were investigated associated with these one dimensional bands, and the Coulomb interaction derived CDW was found to give the highest transition temperature, if the nesting wave number vector was taken as $2\xi_{0+}$ of being shown in Fig. 7. The calculated critical curve is shown in Fig.6, giving a reasonable agreement with the experimental points.

A more detailed experimental work has proceeded.[12)]

3. Electric Conduction and Magnetoresistance of GIC's

3-1 Temperature and Stage Dependence of Resistivity

In many GIC's, the resistivity is known to change with temperature as T^2 instead of the ordinary proportionality to T. A new example of C_6Li[13)] is shown in Fig.8, where T^2-term is leading above ca.100 K. The Table 2 presents the values of coefficients A, B and C in the resistivity formula $\rho = A + BT + CT^2$ for C_8M (M = K, Rb, Cs). The first

TABLE 2. Coefficients A,B,C in the resistivity formula $\rho = A + BT + CT^2$ for C_8M (M = K, Rb, Cs)

intercalant	A [μΩcm]	B [μΩcm/K]	C [μΩcm/K²]	ref.
K	1.86	7.48×10^{-3}	1.536×10^{-4}	14)
K	8.36×10^{-3}	3.4×10^{-3}	9.34×10^{-5}	15)
Rb	9.66×10^{-3}	7.10×10^{-3}	7.63×10^{-5}	15)
Cs	4.06×10^{-2}	7.26×10^{-3}	6.97×10^{-5}	15)

term A is the residual resistivity and depends upon the sample quality. Though the values of B and C of two samples of C_8K are not in agreement with each other, the term CT^2 is several times larger than BT in every specimen in the table. The term BT means the ordinary electron-phonon scattering term, while CT^2 may be due to an electron-electron scattering in a case of poor conductor. In the present case, the factor C could be very small because of a good screening for the electron-electron interaction. Kamimura et al.[17,18)] considered the electron-phonon scattering into two processes: the intra-pocket scattering and the inter-pocket scattering. It is reasonably supposed that GIC'c have, at least partly, cylinder-like Fermi surfaces of which center lines are the zone edges HKH. The inter-pocket scattering means the scattering from a cylinder to the neighboring one, and this scattering probability is enhanced in GIC than the case of pristine graphite because the inter-pocket distance is smaller due to the increase of cylinder diameter. Because a large momentum of phonon is needed for the inter-pocket scattering, the scattering pribability increases more rapidly with temperature than BT of the intra-pocket scattering, so that the resistivity contains the term CT^2 as an approximation.

A similar temperature dependence of resistivity has been reported on acceptor GIC's, e.g., $C_{24}AsF_5$.[16)]

GIC's have such a peculiar dependence of resistivity on the stage number as exhibited in Fig.9.[19)] The room temperature conductivity of alkali metal GIC's is plotted vs. the reciprocal of stage number up to more than 10th stage. The maximum of conductivity appears not at the 1st stage but at the 5 to 7th stage. This behavior was also explained by Kamimura[18)] in a following way. As regards the charge distribution from the carbon layer adjacent to the intercalants to the interior carbon layers, which has been one of the main problems of GIC's, simplified two extreme models have been considered. One is the uniform distribution (UD) from layer to layer along the c-axis, and the other is a metallic sandwich (MS) where the transferred charge is localized on the carbon layer adjacent to the intercalants. The UD model may be crudely applied to the lower stage compounds and the MS model may be a fair approximation for higher stages by virtue of the screening of intercalant charge by the carrier charge. In the UD model, the constancy of the resistivity by the stage variation is obtained as the first approximation, because the $E \propto k$ relation holds for a graphite-like band instead of the ordinary $E \propto k^2$ relation and then the Fermi energy dependences of the carrier concentration and the state density which affects the scattering should compensate with each other. The observed decrease of conductivity with the lowering of stage number is understood by the enhancing of the inter-pocket scattering due to the growing of the pocket diameter.

In the MS model, on the other hand, the conductivity decreases with 1/n, where n is the stage number, since the interior carbon layers have

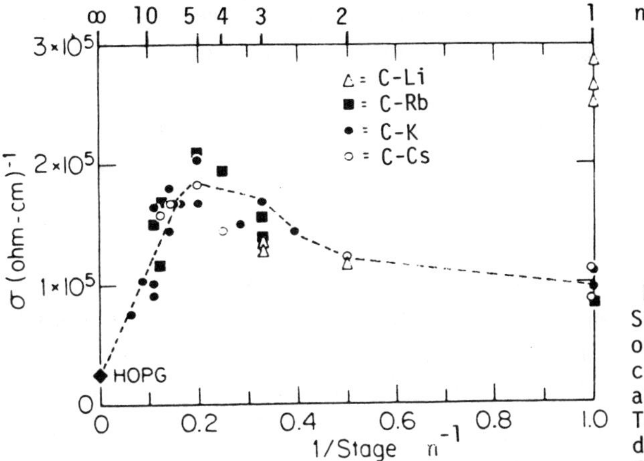

Fig.9
Stage dependence of the electrical conductivity of alkali metal GIC's. The dashed line is drawn by the author for a easy view. (after E.McRae et al.)

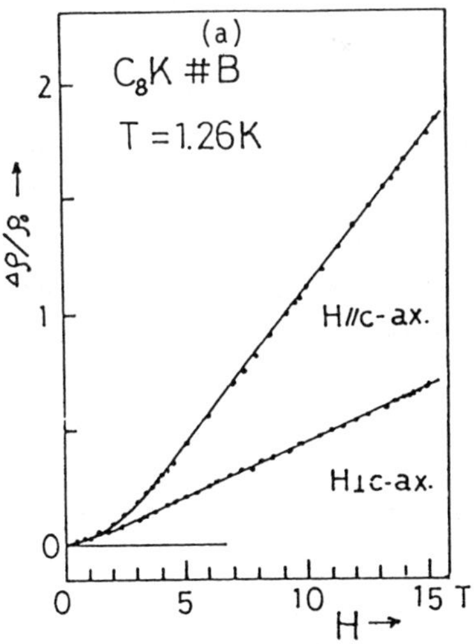

Fig.10
Transverse magnetoresistance of GIC's showing the linear relationship with the magnetic field.
(a) C_8K (after H.Suematsu et al[20])
(b) C_6Li (after R.Inada et al.[21])

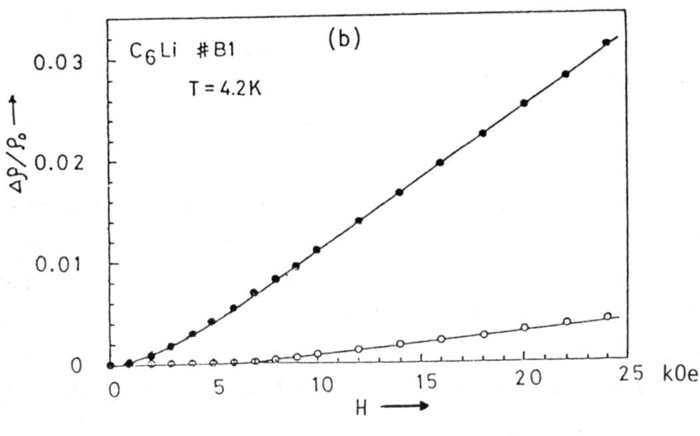

as low conductivity as the pristine graphite one's, and the conductivity is the conductance of the adjacent layer divided by the periodic thickness I_c which is proportional to n.

The remaining problem is the reality of UD model for lower stage compounds except 1st and 2nd ones, as will be discussed in later sections.

3-2 H-Linear Magnetoresistance

The linear relationship between MR and magnetic field intensity was discussed in the section 2-2 on pristine graphite. A similar relationship has been found by the author's group on a fair number of GIC's as exhibited in Fig.10.[20,21] In GIC's, as the Fermi energy is much larger than that of pristine graphite, the magnetic quantum limit should never be attained at the highest field, i.e., 15T in these experimental curves; furthermore, the linearity takes place at such fairly weak fields as a fraction of tesla. Therefore, the explanation for the H-linear MR described in the section 2-2 can not be applied here and the mechanism is left unsolved.

4. Magnetism of Intercalants and of Conduction Carriers

The magnetism of GIC's is one of the most recent and interesting topics in the physics of intercalation. The magnetism consists of two parts; one is originating in the intercalant species and the other is a special paramagnetism of conduction carriers. Interactions between the intercalant spins and the π-electron spins should be also properly taken into account.

4-1 Magnetism of Intercalants

Researches on ferromagnetism or antiferromagnetism have been developed much in Japan from view points of two dimensional magnetism. In practice, C_6Eu and $NiCl_2$- and $CoCl_2$-GIC are extensively investigated.

(1) C_6Eu

Eu and Yb are known to intercalate in graphite as atomic species of rare earth element and behave as donor intercalants.[22] The chemical nature of Eu or Yb resembles the alkali earth element, thus the intercalant ion is Eu^{2+} or Yb^{2+} and two electrons per atom are donated in the graphite π-band, if the full ionization is the case.

Suematsu et al. synthesized C_6Eu with highly oriented pyrolytic graphite (HOPG) and measured the magnetization of the material with a superconducting magnet up to 4 T, clarified that this material is an antiferromagnet having T_N at 40 K,[23] and extended the measurement above 30 T in order to obtain the whole process of magnetization by the joint use of the pulsive high field laboratory of Osaka University.[24] The result for the field parallel to the intercalant plane is shown in Fig. 11. The full line is the experimental curve at 4.2 K and the dashed curve is the theoretical one at 0 K. By the analysis of Suematsu et al., the saturation magnetization above H_{c2} corresponds to the 7 Bohr magneton (μ_B) / Eu^{2+} ion, and the plateau between H_{co} and H_{c1} has the 1/3 height of the final saturation value. Because the magnetization curve for the field perpendicular to the plane, which is shown in the figure as the dotted line, saturates more slowly and has no plateau, spins are thought to be planar (XY-type), so that the spin configurations corresponding to the saturation and the plateau are the ones exhbited in Fig.12(a) and (b), respectively. Furthermore, the magnetic state between 0 and H_{co} is antiferromagnetic corresponding to the Δ-type configuration shown in Fig.12(c). The triangular lattice is so called 'flustrated', so that the Δ-state is easy to change with a small differnce of free energy. The state betweem H_{c1} and H_{c2} corresponds to a canting of spins in order to shift the ferrimagnetic (FI) state of figure (b) to the ferromagnetic (F) state of figure (a).

Thus the every thing seemed clear, but it was found that the FI state could not be stabilized as far as the ordinary two spin interactions were only taken into account. Recently Date et al.[25] obtained an interesting conclusion that the four spin interaction should play the key role. Because C_6Eu is a metal, the exchange interaction between spins was thought to be the RKKY type, i.e., the interaction via conducting π-electron. A simple calculation by the two spin RKKY nearest neighbor interaction J_o, which is shown in Fig.12, for the Δ- and FI-states, gave the same free energy at the field $H_{c1} = 6J_oS/g\mu_B$, but the FI state gave a slightly higher energy in lower fields. This energy order did not change by taking account the second nearest neighbor interaction J_1 or even all of the in-plane remote interactions, and also the inter-plane nearest interaction J'. By taking account the main terms of four spin interactions as follows, the FI state was stabilized in the field range from H_{co} to H_{c1}. The main relevant parts are:

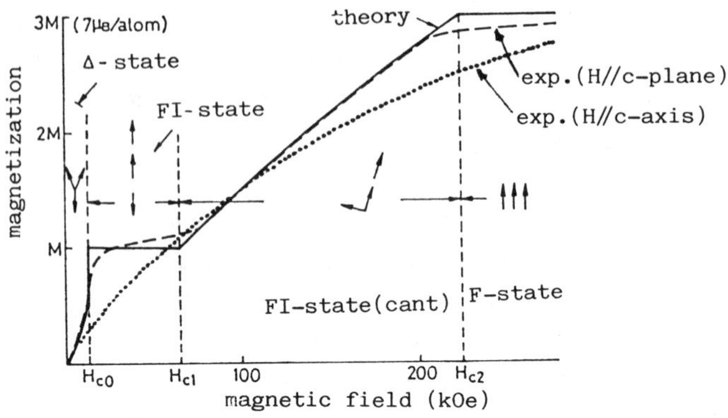

Fig.11
Magnetization curves of C_6Eu at 4.2 K. The dashed line is measured for H parallel to the c-plane and the dotted line for H parallel to the c-axis. The full line is the theoretical curve at 0 K for H parallel to the c-plane.
(after H.Suematsu et al.[24] and M.Date et al.[25])

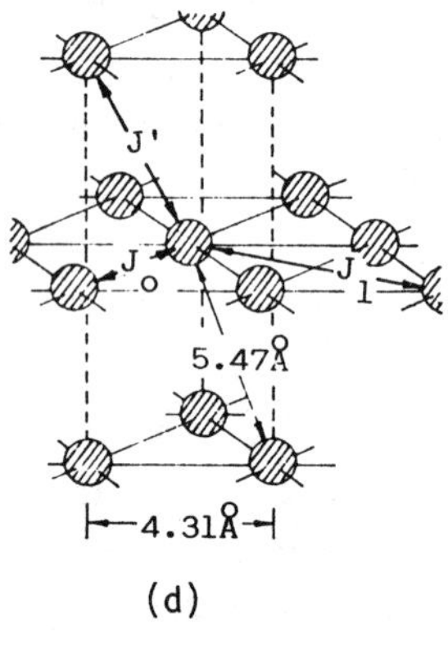

Fig.12
Configurations of Eu^{2+} spins and Eu atoms in C_6Eu.
(after H.Suematsu et al.[24] and M.Date et al.[25])

(a)Ferromagnetic state (saturation magnetization)
(b)Ferrimagnetic state(1/3 plateau)
(c)Antiferromagnetic state(Δ -state)
(d)Configuration of Eu atoms(graphite planes are abbreviated.)

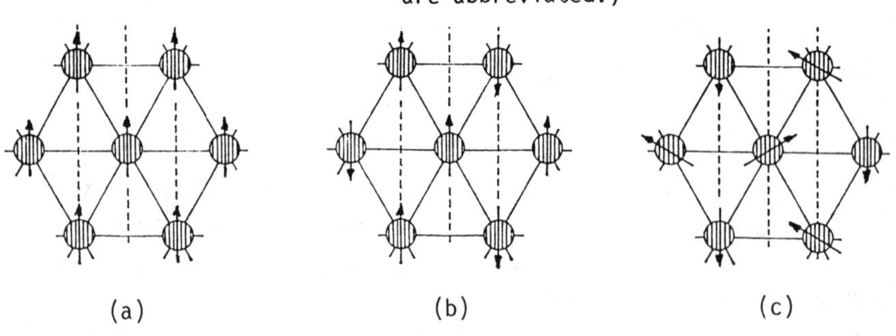

$$\left.\begin{aligned} U_B &= -B(S_iS_j)^2, \qquad B>0 \\ U_K &= K\{(S_iS_j)(S_kS_l) + (S_iS_l)(S_jS_k) - (S_iS_k)(S_jS_l)\},\ K>0 \end{aligned}\right\} \quad (4)$$

where U_B is the biquadratic term and U_K is the four spin nearest ring process. In order to explain the experiment quantitatively, the interaction terms were determined as J_0=0.5 (antiferro.), J_1=0.4 (ferro.) J'=0.1 (ferro.), BS^2=0.02 and KS^2=0.05 (unit is Kelvin). B/J_0 and K/J_0 are so small as 0.004 and 0.009, respectively. Thus the four spin interaction has been found in the RKKY system for the first time.

Akera and Kamimura has developed a non-parameterized calculation on the magnetism of C_6Eu.[25)]

(2) $NiCl_2$- and $CoCl_2$-GIC's

Pioneering work on the magnetic property of the 2nd stage MCl_2-GIC (M = Ni, Co) was done by Karimov et al.[26)]; the results were the followings. (1) Magnetic phase transitions occur succesively at T_{cu} and T_{cl}, where T_{cu} and T_{cl} are 20.2 K and 18.1 K respectively for $NiCl_2$-GIC, and 9.05 K and 8.0 K for $CoCl_2$-GIC. (2) The low temperature phase ($T < T_{cl}$) is ferromagnetic and the intermediate phase ($T_{cl} < T < T_{cu}$) is so peculiar that the spontaneous magnetization M_s is zero but the susceptibility χ diverges. These things seem to correspond to the theoretical prediction for the case of a spin system which has a ferromagnetic planary interaction and a six-fold planary easy axes of magnetization.[27)] If this is the case, the intermediate phase of the two-dimensional XY-spin system has come to be proved by the experiment of $M_s = 0$ and $\chi = \infty$. The ideal system was treated theoretically by Kosterlitz and Thouless[28)] but no practical material had found yet.

Suzuki et al. have investigated the magnetic properties of the same compounds by the use of high quality specimens synthesized from single crystal Kish graphite.[29,30)] They have adopted a high sensitive SQUID magnetometer and have measured M_s and χ at very weak field of less than 3 mOe. The measured curves on $NiCl_2$-GIC is shown in Fig.13. The results are not fully consistent with the former results of Karimov et al. It is seen from the figure that the ordering process has two steps. χ^{aa} increases by lowering the temperature and tends to diverge towards T_{cu} (= 21.3 K) but does not diverge actually and makes a shoulder at T_{cu}, then increases again to make a peak at T_{cl} (= 18.0 K) . M_s is not zero in the intermediate phase ($T_{cl} < T < T_{cu}$) but makes a plateau and vanishes above T_{cu}. Such differnces between the new and old data could

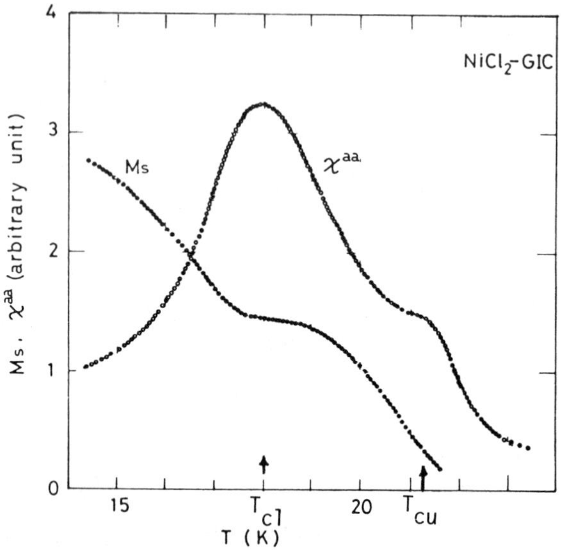

Fig.13
Temperature dependence of magnetic susceptibility χ^{aa} and spontaneous magnetization M_s along the c-plane of $NiCl_2$-GIC single crystal.
(after M.Suzuki et al.[29])

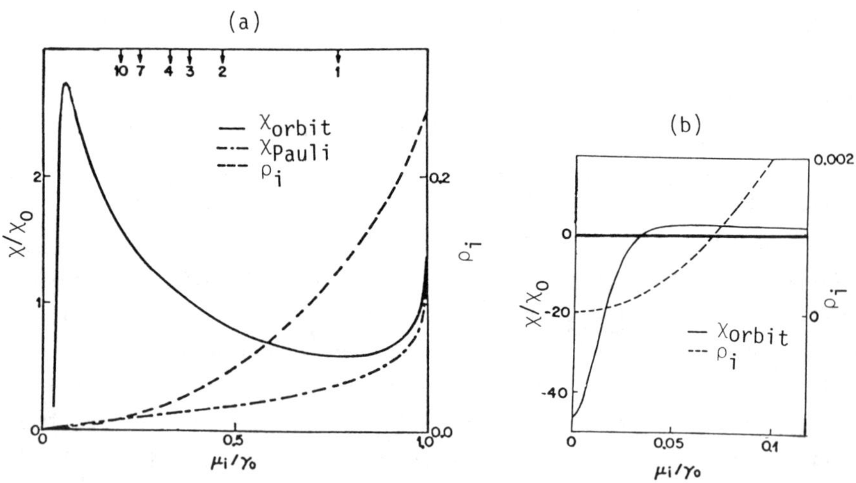

Fig.14
Calculated orbital and Pauli susceptibility of single graphite layer as a function of the Fermi energy μ/γ_0. (γ_0=3.11 eV)
The right hand scale is for ρ_i, the number of electrons per carbon atom in the i-th layer. The top scale is the stage number under the UD model. (b) is the enlarged figure of (a) for the small value of μ/γ_0.
(after F.J.DiSalvo et al.[35])

be ascribed to the difference of magnetic field at which the data was taken, i.e., Karimov et al. had used too large field (1 - 1000 Oe), while Suzuki et al. used as low field as nearly zero. Actually a strong dependence of magnetic property of the intermediate phase on the field strength was proved. Suzuki et al. are apt to think that their intermediate phase should correspond to a pseudo two-dimensional XY-spin system being very near to the Kostarlitz - Thouless phase.

4-2 Susceptibility due to Conduction Carriers

It has been clarified that GIC's have peculiar magnetism due to the orbital motion of conduction carriers, which strongly dependent on the stage number as well as on the charge distribution from a carbon layer to the other along the c-axis direction.

The magnetism of conduction electrons consists of the spin (Pauli) susceptibility, the intraband orbital (Landau-Peierls) susceptibility and interband orbital susceptibility. The latter two were involved in Fukuyama's theory on the orbital magnetism of conducting materials,[31] which succeeded the full explanation of the large diamagnetism of semimetal bismuth. By the theory, the interband transitions dominates in case of bismuth in which a small gap is laid just under the Fermi level. For the case of pristine graphite, McClure derived an expression for the orbital susceptibility χ_{or}^{c} by summing over the Landau levels obtained in a two dimensional model.[32] This calculation had given a large value of observed diamagnetic susceptibility (-250×10^{-6} emu/mol), also by the interband part. Sharma et al. applied Fukuyama's formalism to the three dimensional graphite to obtain a more accurate χ_{or}^{c}.[33] A large diamagnetic value of χ_{or}^{c} is obtained only when the Fermi energy is as small as the value of pristine graphite, 24 meV, while the Fermi energy becomes very large (1-2 eV) for GIC by virtue of the charge transfer. Safran and DiSalvo calculated the magnetic susceptibility of GIC's with the use of the general formula for χ_{or}^{c} derived by Fukuyama[31] and the tight-binding representation for the two-dimensional graphite energy bands.[34] As for the two-dimensional model of GIC, the Fermi energy μ was written as

$$\mu = \mu_i + V_i, \qquad (5)$$

where i is an index specifying the graphite layer, μ_i represents the contribution to the Fermi level from the carriers in that layer, and V_i is the potential at layer i due to the screened intercalant layer, that is to say, the intercalant potential provides a constant shift of the two-dimensional graphite band of each layer. The electron charge density in each layer is thus an adjustable parameter in the calculation. The orbital susceptibililty χ_{or}^{c} is expressed as follows,

$$\chi_{or}^{c}(\varepsilon_0) = \chi_0 \int_0^{\infty} d\varepsilon h(\varepsilon)\left[\frac{1}{2\varepsilon} \frac{\partial}{\partial \varepsilon}\left(\frac{f(\varepsilon-\varepsilon_0) - f(-\varepsilon-\varepsilon_0)}{2\varepsilon}\right)\right], \quad (6)$$

where ε is the energy normalized by the overlapping integral γ_0 appearing in Eq.3, i.e., $\varepsilon = E/\gamma_0$, $\varepsilon_0 = \mu/\gamma_0$, $f(\varepsilon)$ is the Fermi factor and the constant χ_0 is the following quantity.

$$\chi_0 = \left(\frac{e\hbar}{mc}\right)^2 \cdot \frac{1}{2\gamma_0 \Omega_0} \sim 2 \times 10^{-6}, \quad (7)$$

where $\Omega_0 = 1/2 \cdot \sqrt{3} \cdot a^2 c_0$ is the volume of primitive unit cell of a single layer. The factor $h(\varepsilon)$ in the integrant of Eq.6 is the band curvature factor in the relevant graphite model.

The limiting cases of $\varepsilon_0 \ll 1$ and $\varepsilon_0 \simeq 1$ were calculated analytically.

(i) $$\chi_{or}^{c} = -\left(\frac{e\hbar}{mc^*}\right)^2 \frac{1}{8\pi c_0 a^2}\left(\frac{m}{m^*}\right)^2 \frac{1}{k_B T} \operatorname{sech}^2(\mu/2k_B T) + \bar{\chi}, \quad (8),$$

for $\varepsilon \ll 1$, and

(ii) $$\chi_{or}^{c} = \chi_0\left[\frac{1}{8}\left(\frac{m}{m^*}\right)^2 \cdot \rho(\varepsilon_0) + \bar{\rho} \right], \quad (9)$$

for $\varepsilon \simeq 1$. $\bar{\chi}$ and $\bar{\rho}$ are small constants. Eq.8 gives a diamagnetic large susceptibility. The expression agrees with the calculation of McClure[32)] and includes the inter-band effect as the major role. On the other hand, Eq.9 is large and paramagnetic near the saddle point in the density of states. Eq.9 except $\bar{\rho}$ can be derived from a calculation of the Landau-Peierls susceptibility (χ_{LP}^{c}) which yields

$$\chi_{LP}^{c} = -\tilde{\chi}(m_{xx}^* m_{yy}^*)^{-1} \cdot \rho(\varepsilon_0), \quad (10)$$

where $m_{\alpha\beta}^*$ is the effective mass tensor and $\tilde{\chi}$ is a positive constant. Since $m_{xx}^* m_{yy}^* < 0$ near the saddle point, the susceptibility is paramagnetic for $\varepsilon_0 \simeq 1$, and the interband contribution is negligible.

The overall change of susceptibility by the Fermi energy is exhibited in Fig.14. Figure(a) shows the paramagnetic part of which maximum occurs at $\mu_i/\gamma_0 \simeq 0.07$ with the magnitude of $\chi/\chi_0 \simeq 2.7$, while figure(b) shows the transition from paramagnetic to diamagnetic. The diamagnetism at $\mu_i/\gamma_0 = 0$ reaches $\chi/\chi_0 \simeq -46$. The dashed lines in these figures indicate the transferred electron number per carbon atom ρ_i. At the paramagnetic peak, $\rho_i \simeq 0.001$, so that the stoichiometry of $C_{12n}M$ as an example, is $C_{12\times 80}M$, i.e., the 80th stage, if the uniform charge distribution (UD-model in the section 3-1) is assumed. At the top abscissa of Fig.14(a), the stage number for the UD-model is scaled. For the realistic charge distribution, the scale of the stage number should be much different, therefore, the measurement of χ^c_{or} as a function of stage number n is helpful for the determination of the charge distribution. Furthermore, DiSalvo et al. claim that the ratio of χ^c_{or} to the Pauli paramagnetic susceptibility χ_p ($\equiv r$) is better for the aim of determining the charge distribution. In Table 3, experimental values of the susceptibilities are listed. The measurement on C_xK was done by DiSalvo et al.,[35] and the measurement on C_xSbCl_5 has been done by Yoshida et al.[36] In the table, χ_o is the core diamagnetism, χ_p

TABLE 3. Room Temperature Susceptibility χ of Graphite and K- and $SbCl_5$-GIC's in [emu/mole×10^{-6}]

material	$\chi^c(\vec{H}\parallel\vec{c})$	$\chi_\perp(\vec{H}\perp\vec{c})$	$\chi^c-\chi_\perp$	χ_o(core)	χ_p(Pauli)	$\chi^c_{or}/\chi_p\equiv r$	stage	
graphite	-253	-5.2	-248					35)
C_8K	+138	+37.9	100	-52	86 (Ref.37)	1.2	1	35)
$C_{24}K$	+491	+16.4	475	-128	83 (Ref.37)	6.5	2	35)
$C_{36}K$	+825	-9.4	835	-188	95 (Ref.37)	9.7	3	35)
$C_{48}K$	+1150	-36.9	1187	-246	82 (est.)	16.9	4	35)
$C_{24}SbCl_5$	-272	-140.9	-131.0	-234	(35)(est.)	9.4	2	36)
$C_{48}SbCl_5$	-294	-241.6	-52.6	-351	(35)(est.)	12	4	36)
$SbCl_5$		-120.6 (Ref.38)						
$(SbCl_5)^-$		(-420)(est.)						

is the Pauli paramagnetism and χ^c_{or} is the sum of the intraband and interband orbital magnetism which is obtained as $\chi^c - \chi_o - \chi_p$.

The two main features in the table are (i) a large paramagnetic χ^c of C_xK and (2) an increase of r as the stage increases. The orbital χ^c_{or} of C_xSbCl_5 can be paramagnetic if the diamagnetism of $(SbCl_5)^-$ is estimated as ca.-420. and the charge transfer rate is assumed as ca.40%.

5 Superconductivity of Alkali Metal- and Alkali Metal Amalgam- GIC's

Among GIC's, superconductivity was first found by Hannay et al. in 1965, in the first stage alkali metal GIC's (C_8M, M=K,Rb,Cs).[39] The transition temperatures they reported were 0.39 - 0.55 K for C_8K, 0.023 - 0.105 K for C_8Rb, and 0.03 - 0.135 K for C_8Cs. The superconductivity of C_8K was re-discovered with lower T_c by Kobayashi et al.[40] and by the authors group[41,46]. Kobayashi et al. found T_c of 0.08 and 0.125 K for the samples made from graphite powder and grafoil, respectively. The authors group studied more than ten samples made from HOPG to find T_c ranging from 0.128 to 0.198 K. More recently, the second stage potassium-amalgam GIC, C_8KHg, was found to be superconductive at 1.93 K by Alexander et al.[43] and Pendrys et al.[44] Pendrys et al. further found the superconductivity on C_8RbHg at 1.44 K.[44] The author's group re-confirmed these superconducting transitions, and made some detailed investigations.[45,46] The first stage compounds C_4KHg and C_4RbHg were also found to be superconductive by the author's group[45]; however the transition temperatures are 0.6 - 0.9 K lower than those of the second stage compounds. An interest question remains, that is, why the second stage compounds with lower carrier concentrations have higher T_c's than the first stage compounds. Table 4 lists up the superconducting transition temperatures and the anisotropy of the critical magnetic fields of all the superconductive GIC's ever found.

Table 4 Superconductivity of Graphite Intercalation Compounds

Material	Stage	Transition temperature T_c	Anisotropy $H^{\perp}_{c2}$ / $H^{\parallel}_{c2}$	
C_8K	1	0.14 K *	~5	*(0.55 K by Hannay et al.)39)
C_8Rb	1	non > 0.08 K **		**(0.15-0.13 K by Hannay et al.)39)
C_8Cs	1			
C_4KHg	1	0.73 K	~ 10	
C_4RbHg	1	0.99 K	~ 10	
C_8KHg	2	1.90 K	15 - 40	
C_8RbHg	2	1.40 K	20 - 40	
$C_4KTl_{1.5}$	1	2.56 K	~ 10	Ref. 46)
$C_8KTl_{1.5}$	2	1.30 K	—	Ref. 47)

Because the superconductivity of C_8K arose by a combination of non-superconducting elements C and K, the origin of superconductivity is interesting. The intercalant K does not transfer one full electron to the carbon π-band but about 40% of the valence electron remains in the intercalant layer to make a fairly spherical Fermi surface, while the transferred electron makes thick cylinders along KHK edges of the Brillouin zone. Which surface plays the major role for the Cooper pairing? Various considerations do not yet converge. How about the similar problem on $C_{4n}KHg$ or $C_{4n}KTl_{1.5}$? KHg and $KTl_{1.5}$ act as weak donors, so that the charge transfer rate f is smaller than unity. Thus such two kinds of Fermi surfaces should exist as the ones in C_8K. The Cooper pair is likely to asise in the intercalant layer by the following reasons;(1) the measured T_c of KHg itself, 0.94 K, is not far from the T_c values of $C_{4n}KHg$, and (2) the superconducting coherence lengths are estimated as 200 Å and 50 Å for the first and second stage compounds, respectively, of which interlayer distances of the intercalant layers are 10.15 Å and 13.50 Å, which fact seems more understandable if the metal sheet is responsible for the superconductivity.

The coherence length is estimated from the critical field at the absolute zero Kelvin $H_{c2}(0)$ by the following equation; where $H_{c2}(0)$ is determined by the extention of $H_{c2}(T)$-observations to 0 K.

$$\left.\begin{aligned} H^{\parallel}_{c2}(0) &= \frac{\phi_0}{2\,\pi\xi_a^2}, \\ H^{\perp}_{c2}(0) &= \frac{\phi_0}{2\,\pi\xi_a\xi_c}, \end{aligned}\right\} \tag{11}$$

where ξ_a is the coherence length in the planary direction, ξ_c is the coherence length in the c-axis direction and ϕ_0 is the quantum flux, 2.07×10^{-7} G·cm^2. Fig.15 exhibits the stage number dependence of the coherence lengths ξ_a and ξ_c.[45] The 0-th stage means the intercalant amalgam KHg and RbHg themselves. A distinguished feature is that the stage number dependence of ξ_c is almost exponential, while ξ_a has little stage dependence. The cross-over between ξ_c and the c-axis repeat distance may occur between the second and third stages, thus the synthesis of the third stage compound is awaited for the detailed study of the shift of superconductivity from the anisotropic three dimensional type to the two dimensional one.

Another interest is the shift from type-I to type-II superconductivity by the change of external magnetic field away from the c-axis. This was found on C_8K by the author's group. The critical fields H_c, H_{c2}, H_{sc} versus the field angle are plotted in Fig.16. The ranges of field angle for type-I and type-II are divided by the vertical lines. The turning point is determined as H_{c2} is equal to the thermodynamic critical field H_c. The GL(Ginzburg-Landau)parameters $\kappa_{\parallel}$ and $\kappa_{\perp}$, where $\parallel$ and $\perp$ mean

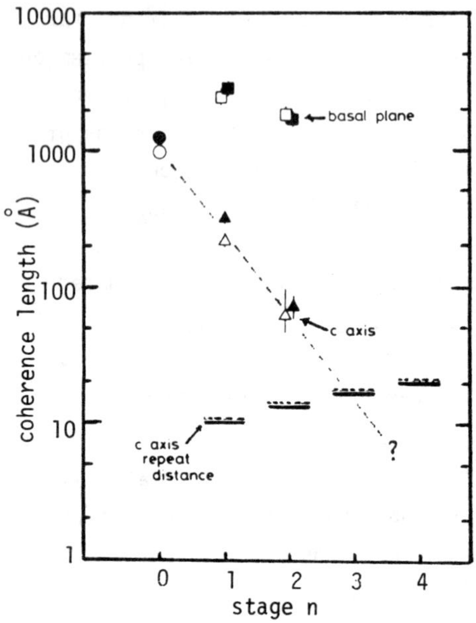

Fig.15
Stage dependence of the basal plane and the c-axis coherence length. stage zero denotes the pristine amalgam. The open and solid symbols represent the KHg and RbHg systems, respectively.
An extrapolation of the stage dependence of the c-axis coherence length is shown by the dashed line. Horizontal lines show the repeat distances of GIC's.

(after Y.Iye et al.[45])

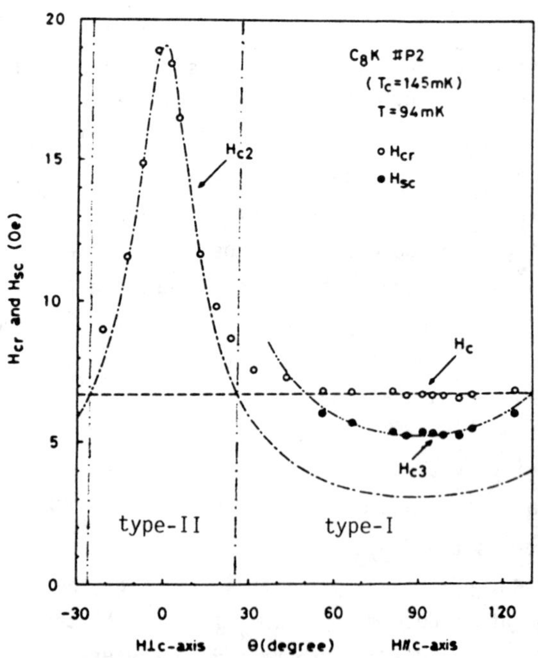

Fig.16
Angular dependences of H_{cr} (H_c or H_{c2}) and H_{sc} of C_8K at 94 mK. The type-I and type-II are separated in angular ranges by the condition $H_{c2} \lesseqgtr H_c$, where H_c is the thermodynamic critical field.

(after Y.Koike et al.[42])

parallel and perpendicular to the c-axis respectively, were determined from the experimental data by the analysis after adopting the anisotropic three dimensional effective mass theory. The obtained values, $\kappa_{\parallel}$= 0.32 and $\kappa_{\perp}$= 2.0,are separated by the criticalvalue, $\kappa = 1/\sqrt{2} = 0.707$ between the type I and the type II.

6 Optical Spectra and Magnetoplasa Wave*

6-1 Optical Spectra of Graphite and GIC's

The optical spectra are determined by the contributions to the dielectric tensor from free carriers (ε_{car}), interband electronc transitions (ε_{int}), phonon excitations (ε_{ph}) and ion cores (ε_{cor}); the last term generally relates to a higher frequency range than the present interest. In the case of graphite and GIC's, the hexagonal symmetry requires two independent tensor components ε_a and ε_c, corresponding to the optical electric field parallel and perpendicular to the layer plane, respectively. In this article, the free carrier and interband contributions are mainly described.

For pure graphite, the role of ε_{int} term is important in the infrared and visible regions of the spectrum, since the valence and conduction bands are near to be degenerate along the zone edge HKH as shown as E_3 band in Fig.2. This E_3 band is further split by the very small spin-orbit interaction ($\sim 2.2 \times 10^{-4}$ eV). Thus the interband transition occurs at a very low frequency and contributes very strongly to ε_{int}. This effect tends to obscure the reflectivity spectrum from which the plasma frequency is identified normally. The plasma frequency for in-plane E vector is estimated as $\sim$0.5 eV corresponding to the room temperature carrier density of 2×10^{-4}/atom and ε_a of 2.8. Another interband effects take place around 0.8 eV as the transitions from the occupied E_2 band to the Fermi level,and from the Fermi level to the empty E_1 band which are shown in Fig.2. In higher energy regions, the inter-π band transition occurs at $\sim$5 eV, and transitions involving σ-band occurs in 13 eV range.

Intercalation with donor or acceptor intercalants much increases ε_{car} at low frequencies due to the large increase of carrier density, e.g., a factor of 800 for the case of C_6Li. At the same time, the inter band effect in lower frequency range is suppressed, because the Fermi level is generally upshifted above the highest band edge of E_3 band for donor compounds and downshifted below the lowest edge of E_3 band for

* The description of §6-1 & 6-2 owes to the review of Profs. Dresselhaus.[48)]

acceptor compounds. Therefore, the optical reflectivity of GIC's exhibits, in general, the characteristic Drude edge at the plasma frequency ω_p. As illustrated in Fig.17, the reflectances of the 1st, 2nd and 3rd stage Rb compounds measured by Guerard et al.[49] show high values for $\omega \lesssim \omega_p$ and decrease rapidly to make minima at the plasma frequencies: $\omega = \omega_p$. Except for the 1st stage, the optical behavior can be described by a single carrier model, and the plasma frequency for the polarization $\vec{E} /\!/ \vec{a}$ is given as follows,

$$\omega_{p,a}^{\ 2} = \frac{4\pi Ne^2}{m_{opt,a} \cdot \varepsilon_{cor,a}} , \tag{12}$$

where $m_{opt,a}$ and $\varepsilon_{cor,a}$ are respectively the in-plane optical effective mass and the in-plane optical core dielectric constant. N means the charge carrier concentration. As for the plasma frequency for the polarization $\vec{E} /\!/ \vec{c}$, the similar expression to the above holds by replacing the subscript a with c. The plasma frequencies are shown in Fig.17 by the virtical arrows, from which N-values in eq.(12) are estimated in such a way that N increases as n decreases.

In Fig.17, the anisotropy of the reflectance is shown too, on C_8K after the work by Zanini and Fischer.[50] The reflectivity curve for the $\vec{E} /\!/ \vec{a}$ polarization is almost the same as the curve for C_8Rb, while the Drude edge for $\vec{E} /\!/ \vec{c}$ polarization is shifted to a lower energy consistent with the lower electrical conductivity for $\vec{E} /\!/ \vec{c}$ and the larger effective mass component m_{zz} relative to m_{xx}. The line shape analyses for both polarizations gave a fair result by adopting two carriers model for the in-plane property and a single carrier for the c-axis property; one of the carriers has a very high mass anisotropy (denoted as 2D carriers) and the other has a lower mass ratio (3D carriers). Such a scheme of carrier types is suggested by the Fermi surface calculation for C_8K by Inoshita et al.,[51] where the 3D - carriers are associated with the quasi-sherical Fermi surface centered at Γ-point and the 2D - carriers are associated with cylindrical surfaces along HKH edges. From their analysis, Zanini and Fischer obtained the mass ratio $m_{zz}/m_{xx} \sim 8$ for the 3D-carrier.

The same kind of measurement as Fig.17 on acceptor compounds C-$SbCl_5$ was done by Eklund et al.[52], as shown in Fig.18. Comparing Figs.17 and 18, the sharp edges for $SbCl_5$-compounds are positioned at lower frequencies than that of Rb-compounds and the edges are almost stage independent. In this case, Eklund et al. considered the interband transitions

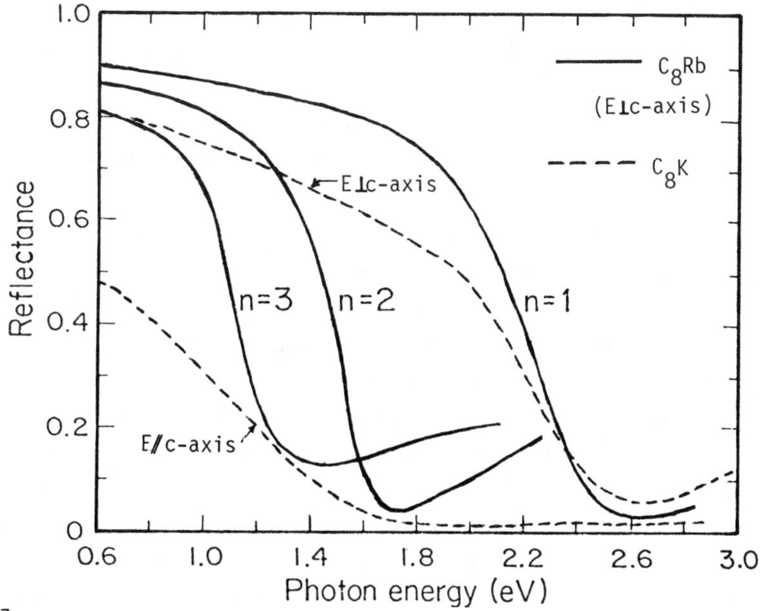

Fig.17
Reflectionspectra of C_8Rb and C_8K. The stage dependence of the plasma frequency is shown for C_8Rb and the dependence of reflectivity on the electric field configuration in relation to the crystal axis is shown for C_8K. (after D.Guerard et al.[49] and M.Zanini et al.[50])

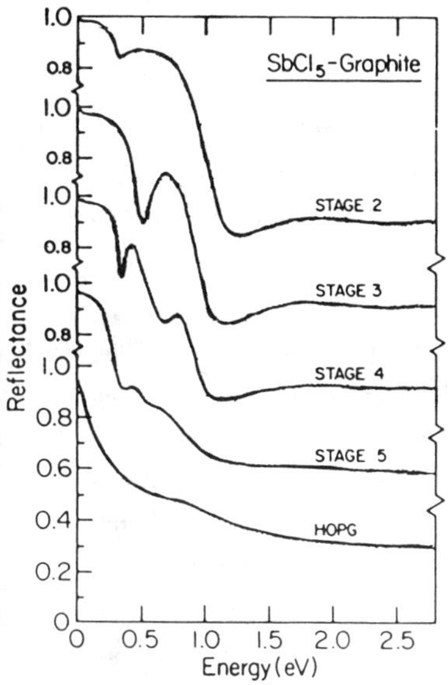

Fig.18
Reflection spectra of $SbCl_5$-GIC's of various stages.
(after P.C.Eklund et al.[51])

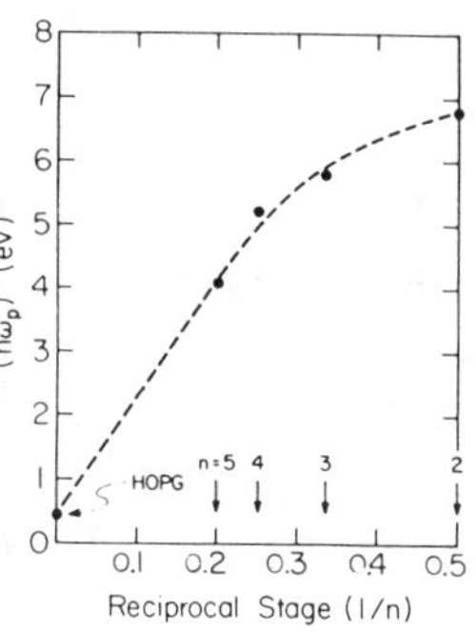

Fig.19
Square of the unscreened plasma frequency versus the reciprocal stage.

of single oscillator types T_{vv} and a threshold type T_{vc} which are designated by the arrows in Fig.18. These interband transitions are generally compatible with the 2D metallic sheet model developed by Blinowski and Rigaux.[53] The unscreened plasma frequencies were estimated by subtracting the interband effect as $\omega_p^2 = 4\pi Ne^2/m^*$ as shown in Fig.19.

6-2 Raman Spectrum

A typical Raman active mode among the lattice vibration modes of graphite is E_{g2} mode of which frequency is 1582 cm^{-1}. This mode is illustrated in the insert of Fig.20. The figure shows the Raman spectra of $FeCl_3$-GIC's reported by Underhill et al.[54] The spectra corresponding to the stages 3 to 11 show the line shapes having two peaks; the α-mode has the same origin as that of pristine graphite and the wave number increases slightly with decreasing of the stage number, and the β-mode is characteristic of intercalation compound. The β-mode comes from the vibration of graphite layer adjacent to the intercalant layer and the α-mode comes from the interior graphite layer(s). The difference of two modes is due to the difference in force constants arising from the differnt environment of the carbon atoms in the graphite bounding and interior layers. The dependence on the reciprocal stage 1/n of the relative intensities of α and β modes suggests that a single bounding layer of either side of the intercalate layer largely screens the intercalate layer from the graphite interior layers.

The frequencies of α and β modes exhibit an upshift with the increase of 1/n for the case of acceptor compounds and a downshift for the donor compounds. These mode stiffening in acceptor compounds and softening in donor compounds are due to lattice strains associated with stage dependent changes in the lattice parameters.

6-3 Magnetoplasma - Helicon - Wave

The electromagnetic wave of which frequency is lower than the plasma frequency can not ordinarily propagate into the metal. Under the static magnetic field, so called magnetoplasma wave propagates throughout the specimen. The magnetoplasma waves are classified in a few proper modes such as cyclotron wave, Alfven wave, Helicon wave etc.by the configuration of $\tilde{E}$ and H, and the strength of magnetic field. Helicon wave is expected to propagate in an uncompensated medium when the propagation vector of the electromagnetic wave is parallel to the static magnetic field, and this type of wave is regarded as the high frequency version

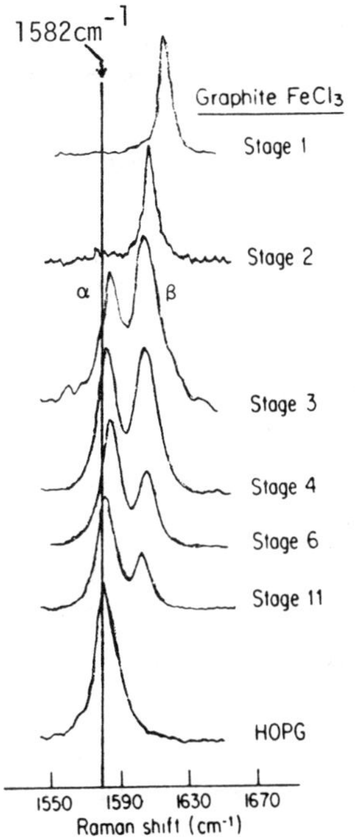

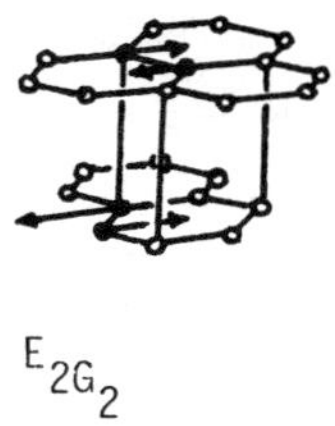

Fig.20
Raman spectra of $FeCl_3$-GIC's.of the stage 1 to 11. The spectra corresponding to the 3-11 stage compounds have the line shape having two peaks of α-mode and β-mode.
The insert shows the crystal structure of graphite and the oscillation mode E_{2g_2} which is responsible for the α-mode in pure graphite. (after C. Underhill et al.[54])

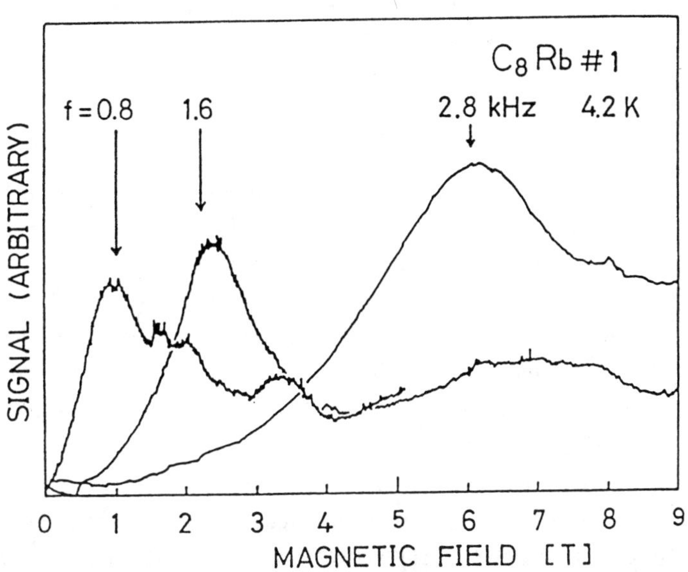

Fig.21
Signal voltage of the Helicon wave detecting coilwound on the specimen of C_8Rb versus the magnetic field, with the propagating frequency as the parameter. The peaks correspond to the Fabri-Perot interference fringe.[55]

of the Hall effect. The view points of magnetoplasma wave as the research object should be (i) search of the possibility of propagation of any magnetoplasma wave and the characterization of the propagated mode, (ii) search of the electronic properties of the media under consideration through the behavior of magnetoplasma wave and (iii) exploratory of new feature of (solid state) plasma. There has been little study of magnetoplasma in layered materials, so that the above view point (i) seems worthwhile to be applied to GIC's. Furthermore, the precise measurement of the Hall coefficient on GIC's by the DC method is not easy due to the problem of getting reliable Hall electrode and uniform primary current, so the Helicon wave may give an alternative mean for getting the Hall coefficient as the view point (ii).

GIC's are metals formed by charge transfer from donor or acceptor intercalants, therefore the carrier are not compensated so that the mode of magnetoplasma should be principally Helicon wave as mentioned above. Trocoidal paths could be easily formed in the c-plane when AC electric field is parallel to the c-plane and the magnetic field is perpendicular to the plane, but the propagation of the wave, which should be along the c-axis, could not be easy because the c-axis conductivity is much less than the in-plane one. In the case of low stage donor type GIC's, the conductivity anisotropy is less than that of pristine graphite, so it is worthwhile to try getting the Helicon propagation in the first or second alkali metal GIC.

The author's group has detected the geometrical interference of Helicon standing wave in parallelepiped specimens of C_8Rb and C_4KHg which were synthesized with the use of HOPG.[55]

The Helicon wave is conventionally detected by observing the interference of waves one of which goes through the specimen and others make multiple reflections by the inner surfaces before passing out. The maximum amplitude is obtained as the following condition holds as the Fabry-Pérot interfernce:

$$d = N \frac{\lambda}{2}, \qquad (13)$$

where λ is the wave length, d is the sample thickness and N is an integer. The sample was wrapped with parafilm and wound by two orthogonal coils, of which one is the exciting coil and the other is the detecting coil. The sample and coil assembly was immersed in liquid helium and the Dewar vessel was inserted in a 10 T superconduction magnet. The signal was

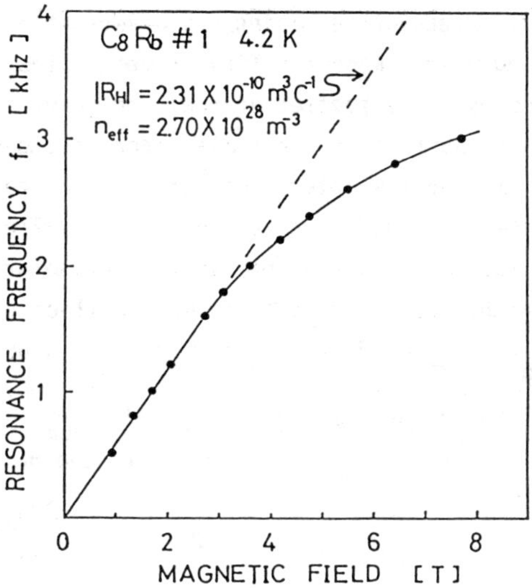

Fig.22
Plot of the Helicon wave frequency versus the magnetic field at which the detecting voltage makes a peak.[55]

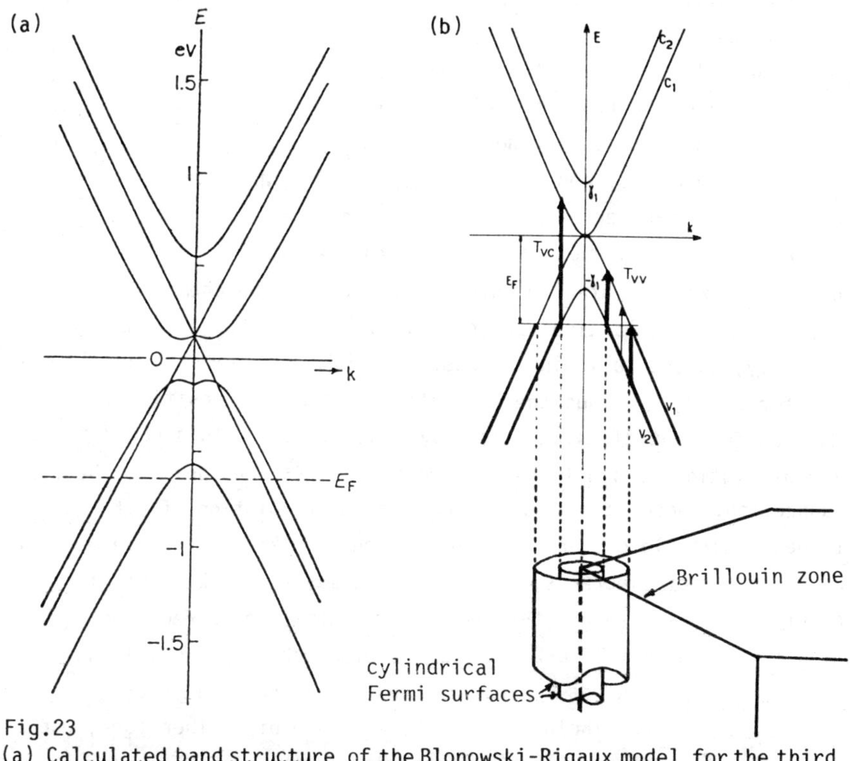

Fig.23
(a) Calculated band structure of the Blonowski-Rigaux model for the third stage acceptor GIC. (b) Schematic illustration for the second stage GIC. (after Blinowski and Rigaux[53])

firstly plotted against the frequency of exciting wave. Among many peaks the peak frquency of depending on the magnetic field strength was chosen and the signal with this frequency was plotted against the field strength as illustrated in Fig.21 as an example. The peaks correspond to the interference fringes of the Fabri-Pérot interferometer. The peak frequencies are plotted versus the field strength in Fig.22. The proportionality of f_r and H in the figure is a characteristic feature of the Helicon wave as seen in the following equation by Merril et al.,[56)]

$$\omega_{r,N} \simeq \left(\frac{\pi N}{d}\right)^2 \rho_H G = \left(\frac{\pi N}{d}\right)^2 \mu_0 G R_H H , \ \rho_H \gg \rho_0, \quad (14)$$

where $\omega_{r,N} = 2\pi f_{r,N}$ is the resonant angular frequency of N-th harmonics, ρ_0 is the resistivity and ρ_H is th Hall resistivity, R_H is the Hall coefficient ($= \rho_0/\mu_0 H$), G is the geometrical constant being almost unity in thin specimens and the sample thickness d is 0.72 mm in the present specimen.

A similar result was obtained on the first stage C_4RbHg compound. The deviation of the data in Fig.22 from the straight line and the same trend on C_4RbHg may mean the dependency of Hall coefficient on magnetic field. From the proportionality of f_r versus H in the measurements, the Hall coefficients were deduced as $|R_H| = 2.3\times10^{-10}\ m^3C^{-1}$ and $2.7\times10^{-10}\ m^3C^{-1}$ for C_8Rb and C_4RbHg, respectively, which values were converted to the carrier densities $2.7\times10^{28}\ m^{-3}$ and $2.3\times10^{28}\ m^{-3}$, respectively, under the assumption of single carrier model. The electron densities are estimated on both compounds from the chemical formulae and the measured lattice spacings, assuming the charge transfer rate f as unity, as the following values: $1.41\times10^{27}\ m^{-3}$ for C_8Rb and $1.56\times10^{27}\ m^{-3}$ for C_4RbHg. Thus the estimation of carrier densities from the Hall coefficient gives unreasonably high values. This could be due to the assumption of single carrier model. As mentioned in the last subsection, Inoshita et al.[51)] made a band calculation of C_8K and obtained a multiple-connected Fermi surface. The Fermi surface consists of two kinds of electrons, one of which comes from alkali metal and nearly sphere-shaped and the others come from transferred electron from alkali metal atom to π-orbit of carbon atom making thick columns along c-axes. These two kinds of Fermi surfaces combine each other to make a complicated surface including many hole-like orbits. Therefore, the transport phenomena should be treated as a multi-carrier conductor. Such an above comparison as the single carrier model gave 20 or 15 times larger values means that an almost complete counter balance between $n_e\mu_e^2$ and $n_h\mu_h^2$ holds, where n_e and n_h are the effective densities and μ_e and μ_h are the mobilities of electron and hole-like orbits, respectively.

7 Problems on Electronic Structure of GIC's

So far, possible electronic structures of GIC's have been mentioned from time to time for the understanding of relevant electrical properties. Namely, two extreme models on the charge carrier distribution along the c-axis: the uniform distribution model and the metallic sandwich (complete screening) model, have been argued in relation to the stage, dependence of conductivity (§3-1), the orbital susceptibility of conduction carriers (§4-1) and the optical spectra (§6-1, 6-2). The famous Fermi surface of C_8K calculated by Kamimura's group[51] and the revision of it by the same group[57] have been used successfully for the qualitative explanation of superconductivity (§5), anisotropy of optical reflectance (§6-1) and Hall coefficient deduced from Helicon wave experiment (§6-3) on the first stage alkali metal GIC's.

More direct cirtifications of the Fermi surface of C_8M (M=K,Rb) calculated by Kamimura's group was done by the author's group by means of de Haas-van Alphen effect measurement.[58-60] The calculated values of de Haas-van Alphen frequencies corresponding to the cross-sectional areas of Fermi surfaces around the points in Brillouin zone, $\tilde{M}$, A(lower band), $\tilde{K}$ (and $\tilde{H}$), A(upper band) are 21, 20, 30 and 11 MOe, respectively, for C_8K. The observed value 28.6 MOe for C_8K could correspond to the calculated value 30 MOe around the $\tilde{K}$ point. In the case of C_8Rb, the observed values 31.3 and 24.6 could simply correspond to the surface sections of $\tilde{K}$ and $\tilde{M}$. The Fermi surface around $\tilde{K}$ (and $\tilde{H}$) is the two dimensional column and the orbit around M is the hole-like orbit, thus the relevant Fermi surface should be fairly realistic.

For the GIC's of the stages 3rd and more, the charge distribution among carbon layers is the most important to be determined. Pietronero et al. estimated the interlayer charge distribution, for the first time, by the use of Thomas-Fermi model.[61] Later, Blinowski and Rigaux[53] proposed a metallic sandwich model as referred in Section 6-1; the sandwich composed of two ending intercalant sheets and n graphite sheets between the intercalant sheets, for the n-th stage compound. Thus the electronic properties are throughly two dimensional. In the n graphite subsystems, the transferred charge distributes most densely in the bounding graphite layers and dilutely in the interior layer due to the screening effect. With the use of SWMcC band model for each graphite layer, the interlayer variation of electrostatic potential was estimated by assuming a proper charge distrubution; thus obtained set of graphite energy bands of all the layers revises the interlayer charge distribution. The process was performed in a self-consistent way. Figure 23 shows the calculated band structure for the third stage acceptor compound. The figure also shows a schematic band structure for the second stage compound. The hole distribution between the bounding and interior layers was calculated as 0.43 : 0.14 for the third stage. The hole Fermi surfaces are composed of n concentric cylinders along the zone edge HKH. For the second stage, the interband transition T_{vv} is single as shown in Fig.23, which was assigned as the single dip of reflection spectrum of the second stage acceptor compounds and this photon energy 0.37 eV was ascribed as the value of band parameter γ_1. From the plasma edge of the reflectance, the carrier density, namely, the charge transfer rate f was obtained as f = 0.27 after a reasonable assumption of γ_0 = 3.12 eV.

This band model seems acceptable as a fair approximation, but the sectional areas of the Fermi surface is at most n kinds. The observed de Haas-van Alphen frequencies are very many in many cases. For example, nine kinds of frequencies were observed in the second stage $SbCl_5$ compound[62], and the in-plane folding of Blinowski-Gigaux band was tried to explain such many kinds of frequencies. This attempt has become unreasonable, because the phase transition from 7×7 commensuration of intercalate layer to glassy state has been found at 210 K with the temperature decrease, by Timp et al.,[63] so that the in-plane band folding has lost the physical meaning if the glassy configuration was smaller than the cyclotron orbits.

More sophisticated band calculations have been reported by Shimomura et al.,[64] Ohno et al.[65] and Holzwarth et al.[66] The most comprehensive calculation is due to Ohno et al. by the use of local density formalism. The calculation has been performed in the self-consistent and non-empirical way on higher stage GIC's by taking account the whole energy band of carbon 1s, σ and π bands. The c-axis charge distribution is tabulated in Table 5. The non-empirical calculation gives the most inhomogeneous distribution. The charge transfer rate f is unity for alkali

Table 5 Comparison of the Theoretical Results of the c-axis Charge Distribution

f=1.0	Self-Consistent			Thomas-Fermi
	Whole Bands	π-Bands only		
layer	Ohno et al.[65]	Safran et al.[67] ε =2	Shimamura et al.[64] ΔE =10 eV	Pietronero et al.[61] ε =5.5
(n=3)				
1	0.493	0.438	0.478	0.420
2	0.013	0.124	0.045	0.160
(n=4)				
1	0.487	0.425	0.472	0.400
2	0.013	0.075	0.028	0.100
(n=5)				
1	0.485	0.418	0.470	-
2	0.013	0.065	0.025	
3	0.004	0.034	0.010	
(n=6)				
1	0.485	0.417	0.469	-
2	0.012	0.059	0.024	
3	0.003	0.024	0.007	
f=0.3				
(n=3)				
1	0.447	0.411	-	0.395
2	0.105	0.177	-	0.210
(n=4)				
1	0.428	0.380	-	0.360
2	0.072	0.120	-	0.140

ε is the dielectric constant and ΔE is the effective excitation energy of carbon $2p_z$ orbitals for polarization, as the parameters. This table is copied from the paper by Ohno et al.[65]

metal GIC's and 0.3 for acceptor GIC's, $C_{12n}SbCl_5$ as a typical example. This value of charge distribution could be the guide for experimental searches. So far, the most direct experimental approach is the ^{13}C NMR. Recently, Kume et al.[68] performed ^{13}C NMR measurement on pristine graphite and the stage 2, 3, 4 and 6 K-GIC's. The resonances have been taken on the HOPG-based bulk samples and higher resolutions have been obtained on the NMR spectrum compared to the previous works. The relaxation time T_1 has been measured as well as the spectrum on every resonance frequency and the T_1T has been found to be constant. This fact means the mechanism of relaxation is dominantly metallic, so the state density N(E) of relavant band can be estimated if the proportional constant between N(E) and T_1T is assumed as common. The obtained ratio of N(E)'s of the layers 1,2 for the stage 3 and 1,2,3 for the stage 6 are respectively, 0.407: 0.185 and 0.412:0.058:0.030. The inhomogeneity is milder than the above described theoretical guiding values. The discrepancy should be removed by more detailed considerations on both of theoretical and experimental points of view. The quantitative understanding of every electronic property of high stage GIC's can not be attained without determining the charge distribution along c-axis.

Acknowledgement

The author thanks Professor Y.Uemura for many discussions and encouragements on the author's works on graphite and related substances in many years. He is also grateful to Professor H.Kamimura for stimulating discussions on works on graphite intercalation compounds. The works of author's group which are cited in this article were supported by Grant-in-Aid for Scientic Research from the Ministry of Education, Research and Culture.

References

1) M.Tsukada, K.Nakao, Y.Uemura and S.Nagai: J. Phys.Soc.Jpn.,32(1972)54.
2) J.C.Slonczewski and P.R.Weiss: Phys.Rev.,109(1958)272.
3) J.W.McClure: Phys.Rev.,108(1957)612;119(1960)606.
4) H.Nagayoshi, M.Tsukada, K.Nakao andY.Uemura: J.Phys.Soc.Jpn.,35(1973) 396.
5) H.Nagayoshi,K.Nakao and Y.Uemura: Solid State Commun.,18(1976)225.
6) J.W.McClure and W.J.Spry: Phys.Rev.,165(1968)809.
7) K.Sugihara and J.A.Woollam: J.Phys.Soc.Jpn.,45(1978)1891.
8) S.Tanuma, Y.Ōnuki, R.Inada, A.Furukawa, O.Takahashi and Y.Iye: Physics in High Magnetic Fields, Springer Verlag Solid State Science 24 ed. S.Chikazumi and N.Miura (Springer,Berlin,1981) p.316.
9) S.Tanuma, A.Furukawa and Y.Iye: Extended Abstract of 15th Biennial Conf. on Carbon, Philadelphia, 1981, ed. W.C. Forsman (Univ. of Pennsylvania Press, 1981) p.16.
10) Y Iye, P.M.Tedrow, G.Timp, M.Shayegan, M.S.Dresselhaus, G.Dreselhaus A.Furukawa and S.Tanuma: Phys. Rev. B25 (1982)5478.
11) D.Yoshioka and H.Fukuyama: J.Phys.Soc.Jpn.,50(1981)725.
12) K.Nakamura, T. Osada, G.Kido, N.Miura and S.Tanuma: in press in J Phys. Soc. Jpn.
13) Y.Koike, A.Furukawa and S.Tanuma: unpublished.
14) H.Suematsu,K.Higuchi and S.Tanuma Physica, 99B(1980)420.

15) M.E.Potter, W.D.Johnson and J.E.Fischer: Extended Abstract of 14th Biennial Conf. on Carbon (Pennsylvania State Univ, 1979) p.300.
16) C.Zeller, L.A.Pendry and F.L.Fogel: J.MAt.Sci.,14(1979)2241.
17) H.Kamimura, K.Nakao, T.Ohno and T.Inoshita: Physica,99B(1980)401.
18) H.Kamimura: Physica,105B(1981)501.
19) E.McRae, D.Billaud, J.F.Mareche and A.Herold: Physica,99B(1980)489.
20) H.Suematsu, K.Higuchi and S.Tanuma: J.Phys.Soc.Jpn.,48(1980)1541.
21) R.Inada, H.Suematsu, K.Higuchi, A.Furukawa, Y.Koike and S.Tanuma: unpublished.
22) D.Guerard and A.Herold: C.R.Acad.Sc.Paris,Serie C 281(1975)929.
23) H.Suematsu, K.Ohmatsu and Y.Yoshizaki: Sol.St.Commun.,38(1981)1103.
24) H.Suematsu, K.Ohmatsu, K.Sugiyama andM.Date: Sol.St.Commun.,40(1981) 241.
25) M.Date, T.Sakakibara, K.Sugiyama and H.Suematsu: Proc.Int'l Symp.on High Field Magnetism,Ohsaka,Japan(1982).
26) Yu S.Karimov and Yu N Novikov: Sov.Phys.JETP Let.,19(1974)547.
27) J.V.Jose, L.P.Kadanov, S.Kirkpatrik and D.R.Nelson: Phys.Rev.,B16 (1977)1217.
28) J.M.Kosterlitz and D.J.Thouless: J.Phys.C,6(1973)1181.
29) M.Suzuki, H.Ikeda, M.Murakami, M.Matsuura, H.Suematsu, R.Nishitani and R.Yoshizaki: J.Mag.Mag.Mater.,31-34(1983)1173.
30) Y.Murakami, M.Matsuura, M.Suzuki and H.Ikeda: J.Mag.Mag.Mater.,31-34 (1983)1171
31) H.Fukuyama: Prog.Theor.Phys.,45(1971)704.
32) J.W.McClure: Phys.Rev.,104(1956)666.
33) M.P.Sharma, L.G.Johnson and J.W.McClure: Phys.Rev..B9(1974)2467.
34) S.A.Safran and F.J.DiSalvo: Phys.Rev.,B20(1979)4889.
35) F.J.DiSalvo, S.A.Safran, R.C.Haddon, J.V.Waszczak and J.E.Fischer: Phys.Rev.,B20(1979)4883.
36) A.Herold: Bull.Soc.Chim.France,3(1955)999.
37) Y.Yoshida, Y.Saito, Y.Iye and S.Tanuma: unpublished.
38) O.Pascal: Comt.Rend.,152(1912)863;218(1944)57.
39) N.B.Hannay, T.H Geball, B.T.Matthias,K.Andre, P.Schmidt and D.MacNair: Phys.Rev.Let.,14(1965)225.
40) M.Kobayashi and I.Tsujikawa: J.Phys.Soc.Jpn.,46(1979)1945.
41) Y.Koike, H.Suematsu, K.Higuchi and S.Tanuma: Sol.St.Commun.,27(1978) 623.
42) Y.Koike, S.Tanuma, H.Suematsu and K.Higuchi: J.Phys.Chem.Sol.,41

(1980)1111.

43) M.G.Alexander, D.P.Goshorn, D.Guerard, P.Lagrange, M.El Makrini and G.Onn: Synth.Met,2(1980)203.

44) L.A.Pendrys, R.Wachnik, F.L.Fogel, P.Lagrange, G.Firdin, M.El Makrini and A.Herold: Sol.St.Commun.,38(1981)677.

45) Y.Iye and S.Tanuma: Phys.Rev.,B25(1982)4583.

46) Y.Iye and S.Tanuma: Synth.Met.,5(1983)257.

47) F.L.Vogel, R.Wechnik and L.Pendrys: Physics of Intercalation Compounds, ed. L.Pietronero and Tosatti, Springer Verlag Solid State Science 38(1981),Berlin,P.288.

48) M.S.Dresselhaus and G.Dresselhaus: Adv. in Phys.,30(1981)139.

49) D.Guerard, G.M.T.Foley, M.Zanini and J.Fischer: Nuov.Cim.,38(1977) 419.

50) M.Zanini and J.E.Fischer: Mat.Sci.Eng.,31(1977)169.

51) T.Inoshita, K.Nakao and H.Kamimura: J.Phys.Soc.Jpn.,43(1977);ibid. 44(1977) 689(errata).

52) P.C.Eklund, D.S.Smith and V.R.K.Murthy: Synth.Met.,3(1981)111.

53) J.Blinowski andC.Rigaux: Synth.Met.,2(1980)297.

54) C.Underhill, S.Y.Leung, G.Dresselhaus and M.S.Dresselhaus: Sol.St. Commun.,29(1979)769.

55) S.Tanuma, I.Oguro and R.Inada: to be published in Synth.Met.(Proc. 3rd Int'l Conf. on GIC's.)

56) J.E.Merril, M.T.Taylor and J.M.Goodman: Phys.Rev.,131(1963)2499.

57) T.Ohno, K.Nakao and H.Kamamura. J. Phys. Soc. Jpn. 47(1979)1125.

58) K.Higuchi, H.Suematsu and S.tanuma: J.Phys.Soc.Jpn.,48(1980)1532.

59) S.Tanuma, Y.Iye O.Takahashi and Y.Koike: Synth.Met.2(1980)341.

60) S.Tanuma: Physica, 165B(1981)486.

61) L.Pietronero, S.Strassler, H.K.Zeller and M.J.Rice: Phys.Rev.Let., 41(1978)763.

62) S.Tanuma, O.Takahashi and Y.Iye: Physics of Intercalation Compounds, ed. L.Pietronero and E.Tosatti,Springer Verlag Solid State Science 38(1981),Berlin,p.90.

63) G.Timp, M.S.Dresselhaus, L.Salamanca-Riba, A.Erbil, L.W.Hobbs, G.Dresselhaus, P.C.Eklund and Y.Iye: Phys.Rev.,B26(1982)2323.

64) S.Shimomura and A.Morita: J.Phys.Soc.Jpn.,51(1982)502.

65) T.Ohno and H.Kamimura: J.Phys.Soc.Jpn.,52(1983)223.

66) N.A.W.Holzwarth,S.G.Louis and S.Rabii:Phys. Rev. Let.,47(1981)1318.

67) S.A.Safran and D.R.Hamman: Phys.Rev.,B23(1981)565.
68) K.Kume, Y.Maniwa, H.Suematsu and S.Tanuma: to be published in Sol. St.Commun.

THEORY OF ELECTRONIC STRUCTURE AND CHEMISORPTION OF SEMICONDUCTOR SURFACES

Masaru Tsukada

Department of Physics, Faculty of Science,
University of Tokyo,
7-3-1 Hongo, Bunkyo-ku, Tokyo 113

ABSTRACT

This article reviews recent development of theoretical studies on semiconductor surfaces, which established close correlation between the electronic structure and atomic arrangement of surfaces. These studies provide crucial informations on electronic, atomic and chemical properties of the surfaces by the first-principles electronic structure theory. The topics are selected from our recent works based on the self-consistent $X\alpha$ calculations. They include 1) electronic structure of vacancy and chemisorption bond on Si(111) surface, and 2) peculiar electronic structure of the polar surfaces and its significance to the chemical reactivity.

1. Introduction

Thanks to recent progress in basic experimental and theoretical works, understanding of solid surfaces in atomic scale has been considerably increased. In particular many newly developed experimental methods such as low energy ion scattering spectroscopy (ISS), low energy electron diffraction (LEED), reflected high energy electron diffraction (RHEED), molecular or atomic beam scattering (MBS), high resolution electron energy loss spectroscopy (ELS), ultra-violet photoemission spectroscopy (UPS) provide reliable accurate data for the atomic and electronic structure of solid surfaces.[1)]

However, through close interplay between the atomic arrangement and the electronic structure, solid surfaces are much more complicated systems than the bulk of crystals. For example, surface atomic structures are not selfevident as in the bulk. Drastic reconstruction structures are usually observed on surfaces of covalent semiconductors and certain transition metals and noble metals. Elucidation of the mechanism of molecular and electronic processes on surfaces is crucially important in various fields of applications. Such processes are naturally governed by the atomic and the electronic structure of the surface.[2)]

It is often the case that a complete understanding of the atomic and the electronic structure of the surface cannot be obtained only by the experimental informations. The surface electronic structure calculated theoretically is quite useful not only to provide complementary informations, but also to give a unifying picture of surface atomic and electronic structure. In particular non-empirical theory of electronic structure is important, because the reliability of the semi-empirical theory is questionable for the surface. For the study of molecular processes on solid surfaces, the Born-Oppenheimer adiabatic potential energy surface is the clue to the elucidation of the mechanism, although at the present stage it might be a formidable task to calculate it non-empirically.[3)] Owing to the development of the theoretical methods and the modern computer systems, one can at least obtain reliable picture of the surface electronic structure for any "assumed" atomic arrangement. Close correlation between the surface atomic structure and the electronic structure has been revealed, which is useful for the selection of the proper structure models and for the elucidation of microscopic properties of the surfaces.

In the present article, we review the non-empirical electronic structure theories based on the local electron density.[4)] They include the local density functional (LDF) method,[4)] $X\alpha$ method[5)] and self-consistent pseudo-potential method.[6,7)] Two problems of the semiconductor surfaces are discussed mainly based on our recent works.[8-15)] They provide the typical examples in which the surface electronic states caused by special surface structure take an important part. One of them is the reconstruction of covalent semiconductors,[8-12)] especially the atomic arrangement and the electronic states of the Si(111)7×7 superlattice structure. The other problem is the peculiar electronic structure and the related properties of the polar compound surfaces.[13-15)]

2. Electronic Structure Theories Based on the Local Electron Density

Semiempirical theories of electronic structure are sometimes unreliable for solid surfaces. One of the reasons is that potential and electron density change drastically as going from the bulk to the surface. Thus the parametrization of the model which is effective in the bulk is not necessarily valid in the surface. The non-empirical method should be used to obtain reliable results of the surface electronic structure. The powerful methods which can be performed without enormous computational task are based on the effective exchange-correlation potential derived by the local electron density.[4] Among them, there are various $X\alpha$ methods,[5,16] the local density functional method[4,17] and the self-consistent empirical or first-principle pseudopotential method.

In these method, one electron wave function ψ_λ is determined by the following equation,

$$\{-\frac{\hbar^2}{2m}\nabla^2 + V_{core}(x) + \int \frac{\rho(x')}{|x-x'|}dx' + V_{xc}(\rho(x))\}\,\psi_\lambda(x)$$

$$= E_\lambda\psi_\lambda(x), \tag{2.1}$$

where the electron density $\rho(x)$ is self-consistently determined as

$$\rho(x) = \sum_{\lambda(E_\lambda \leq E_F)} |\psi_\lambda(x)|^2. \tag{2.2}$$

The exchange-correlation potential V_{xc} in eq.(2.1) is evaluated as

$$V_{xc}(\rho) = \frac{d}{d\rho}\rho\,\varepsilon_{xc}(\rho) \tag{2.3}$$

where $\varepsilon_{xc}(\rho)$ is the exchange-correlation energy per electron for the uniform electron system with the density ρ. Based on many-electron theories of uniform system, some different forms of $\varepsilon_{xc}(\rho)$ have been suggested, although significant differences of the results are not caused. They are, for example,

$$\varepsilon_{xc}(\rho) = -\frac{3}{4}(3\rho/\pi)^{1/3} - \frac{0.056\,\rho^{1/3}}{0.079+\rho^{1/3}} \quad \text{(Wigner)}^{18)} \tag{2.4a}$$

$$\varepsilon_{xc}(\rho) = -\frac{3}{4}(3\rho/\pi)^{1/3} - 0.0155\,\ln(3/4\pi\rho)^{1/3} - 0.0575 \tag{2.4b}$$

(Pines & Nozieres)[19]

in atomic unit. In the $X\alpha$ method,[5] one assumes the form

$$\varepsilon_{xc}(\rho) = -\frac{9\alpha}{8}\left(\frac{3\rho}{\pi}\right)^{1/3}, \tag{2.5}$$

where the value of α should be taken in the range 2/3 $\sim$ 0.7. The term $V_{core}(x)$ in eq.(2.1) is the potential due to the nucleus

$$V_{core}(x) = -\sum_{\nu} \frac{Z_\nu e^2}{|x-R_\nu|}, \tag{2.6}$$

where Z_ν, R_ν is the charge and the position of the ν-th nucleus.

For the case of the self-consistent empirical pseudo-potential method,[6] the self-consistent calculation of the electron density are performed only for the valence electrons. And the ion core pseudo-potential, $V_p(x)$ is substituted for the neucleus potential $V_c(x)$. $V_p(x)$ is determined to reproduce the band structure of the bulk. There have been proposed several forms of $V_p(x)$. For example Appelbaum and Hamann[20] used the analytical expression for Si^{4+};

$$V_p(q) = a_1\left[\frac{1}{q^2} + a_2 + a_3 q^2\right]\exp(-a_4 q^2) \tag{2.7}$$

where a_i (i=1,$\cdots$4) are the parameters to reproduce the bulk band. Corresponding to this weak pseudo-potential, the valence wave function is obtained by the plane wave expansion.

For more sophisticated calculations, say for systems including transition-metals,[7] the self-consistent first-principle pseudo-potential is used. This is so determined to reproduce the valence electronic states of free atom. They are nonlocal, since the electron feels different potential depending on its angular momentum around the nucleus. The wave functions are often expanded by the mixed basis of plane waves and localized atomic orbitals.

Since it is quite laborious to solve the Schrödinger equation (2.1) for the real semi-infinite system, some simple models of the surface are usually introduced. The slab model is often used for the semi-conductor surfaces,[6] because it can describe the energy dispersion of shallow surface states. For the investigations of chemisorption or the localized states due to surface defects, cluster models of the surface are effective.[9-16] This is because such perturbations of the surface are confined within narrow spatial regions.

DV-Xα method is a practical non-empirical approach for the calculations of the electronic structure.[16] Since we will discuss the results obtained by this method in the following sections, its numerical procedure is explained below. The eigen functions of the clusters, $\psi_\lambda(x)$ are expanded by the linear combinations of atomic orbitals $\chi_i(x)$,

$$\psi_\lambda(x) = \sum_i C_{i\lambda}\chi_i(x). \tag{2.8}$$

The atomic orbitals $\chi_i(x)$ are numerically obtained as the eigen functions of "atoms in cluster". The coefficients $C_{i\lambda}$'s are determined by usual variational procedure as the eigen vectors of

$$\sum_j H_{ij}C_{j\lambda} = E_\lambda \sum_j S_{ij}C_{j\lambda} \,. \tag{2.9}$$

The special feature of the DV-Xα method exists in the numerical procedures to evaluate the matrix elements H_{ij} of the Hamiltonian and those of the overlap integrals, S_{ij}. Namely they are calculated by the random sampling method as

$$H_{ij} = \sum_k \omega(x_k)\chi_i^*(x_k)h(x_k)\chi_j(x_k) \tag{2.10}$$

$$S_{ij} = \sum_k \omega(x_k)\chi_i^*(x_k)\chi_j(x_k), \tag{2.11}$$

where x_k is one of the random sampling points, $\omega(x_k)$ is the statistical weight of the point x_k and $h(x_k)$ is the one-electron Hamiltonian.[13,16]

3. Reconstructions and Chemisorptions of Semiconductor Surfaces

3.1 Mechanism of simple reconstructions

A variety of reconstructed superlattice structures are observed on covalent semiconductor surfaces.[1,21] The structures change depending on materials, crystallographic direction of surface, methods of preparation and temperatures. On the surface of semiconductors, the covalent bond tying up neighboring atoms together in the bulk is broken, leaving the dangling bond, i.e. the non-bonding orbital on the top atom. Although the detailed atomic structure and mechanism of most of surface reconstructions remain to be solved yet, their origin is believed to be the heighly raised energy of the dangling bond.

Namely reconstruction takes place for many cases to rebond the dangling bond each other or to stabilize the energy of the dangling bond occupied by electron. For the polar surfaces of compound crystals, such reconstructions often take place as recover the electric neutrality of the surface.[22] This will be discussed in section 4.

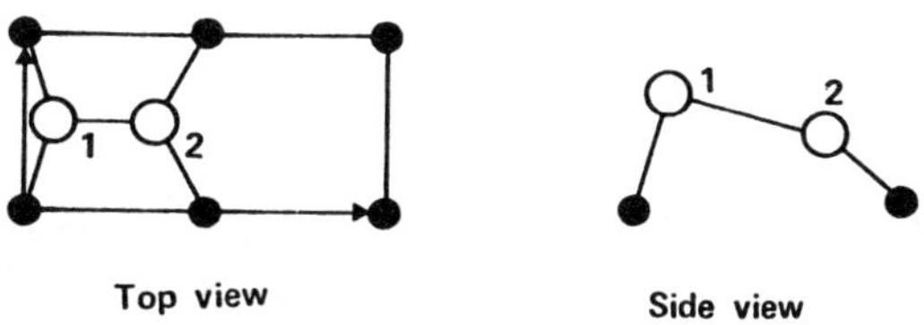

FIG.1. Top and side views of the asymmetric dimer model. 1 and 2 denote the raised and lowered Si atoms in the dimer.

The well known example of the reconstruction is the dimer structure of the Si(100) 2×1 surface.[1,21,23] In this structure, each two nearest neighboring top Si atoms come close together to form a dimer. Then the dangling bonds of the two atoms nearly parallel to the surface are changed to the σ bonding orbital of the dimer. By more detailed LEED observation and the He atom scattering, the Si(100) surface has been now revealed to be C(4×2) structure.[21] This structure was clarified by Chadi[21,23] based on the asymmetric dimer model, which includes the charge transfer between the two atoms in the dimer.[21] The negative atom in the dimer ("1" in Fig. 1) is shifted towards the vacuum, making the vertical angle of the three back bonds sharper (Fig. 1). The sharpening of the vertical angle lowers the energy of the dangling bond, which accepted the electron from the counter ion in the dimer. On the contrary, the more obtusely directed back bonds of the positive Si atom in the dimer ("2" in Fig. 2) raises the energy of the dangling bond. This is just the mechanism to stabilize the asymmetric dimer compared with the symmetric one.

The same mechanism is working for the reconstruction structure of (110) cleavage surface of III-V compound semiconductors.[1,24] The reconstructed surface has the same periodicity as in the corresponding plane in the bulk. But, as shown in Fig. 2, the plane of the zig-zag line connecting the neighboring cation and anion is rotated by some angle θ from the original (110) layer plane. As the result the

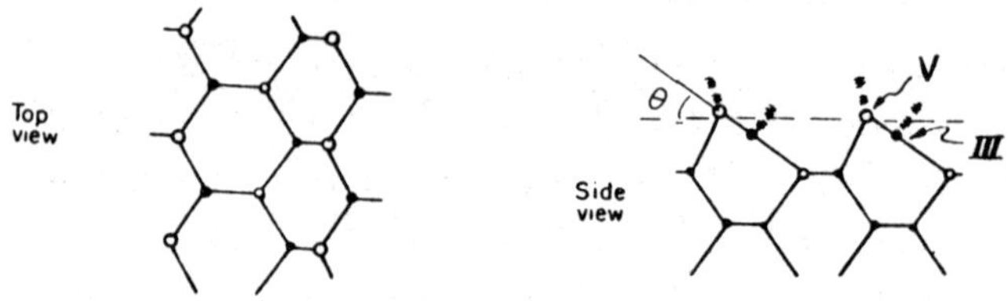

FIG. 2 The structure of the relaxed (110) surface of III-V compounds.

vertical angle made of the back bonds becomes sharper for the anion and that for the cation becomes more obtuse. Consequently some part of the electron density is transferred from the cation dangling bond to the anion dangling bond, because of the raising of the former energy and the lowering of the latter energy. The calculated energy position of the dangling bonds for the reconstructed structure are removed from the gap region, in good correspondence with experiments.[1,24)]

3.2 Models of Si(111) 7×7 structure

The atomic structure and the mechanism of the long range superlattice structure such as Si(111) 7×7 or Ge(111) 2×8 have been the subject of the hot controversy and still remain unresolved difficult problem. Numerous structure models of Si(111) 7×7 surface have been proposed based on various experiments such as LEED, RHEED, UPS, He scattering and ISS, as well as theoretical speculations. They include the vacancy model by Lander-Morrison,[25)] that by Ino,[26)] the ring-like buckled model by Chadi,[27)] the step array model by Pandey,[28)] the adatom model by Harrison.[29)] For a model to be unambiguously established, it must explain all kinds of experiments, concerning both on atomic structure and electronic structure. And such model should be also consistent with the behavior of structure change and that of chemisorption.

For the very complicated system as Si(111) 7×7 structure, it is almost impossible to determine the stable atomic geometry from the first-principles calculations. At the present stage, however, it might be useful to clarify theoretically the characteristics in the electronic structure, which are specific to assumed surface structures. By comparing them with experimental informations on electronic structure, for example by UPS, ELS, or work function measurements, one can at least select proper structure models.

3.3 Electronic structure of Si(111) surface with vacancies

The vacancy model of Si(111) 7×7 super-lattice is originally proposed by Lander-Morrison.[25] By detailed analysis of RHEED intensity distribution pattern as well as LEED I-V relation, Ino proposed recently a new vacancy model.[26] In this model three vacancies forming a regular triangle with the side length 2a (a=3.84Å is the lattice constant of the unreconstracted surface) are included in the unit cell of the super-lattice (see Fig. 3). Vacancy model seems consistent with the microscopic phase transition process between 7×7 and 1×1 structure observed with the reflection electron microscope by Osakabe, Yagi and Honjo.[30] The step array model by Pandey[28] and the ring-like buckle model by Chadi[27] would consistent neither with the microscopic phase transition process[30] nor the intensity pattern of RHEED.[26]

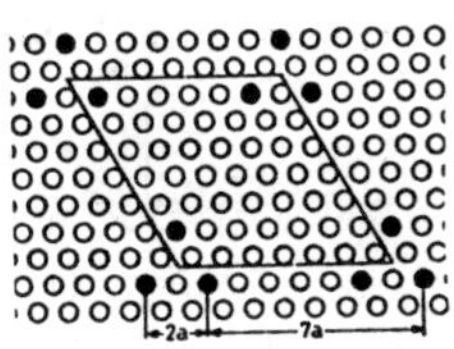

FIG. 3. Ino's vacancy model of Si(111) 7×7 surface.

It is interesting to see what characteristics in the electronic structure are introduced by the vacancy. Nagayoshi and Tsukada[8] performed selfconsistent pseudopotential calculations for the Si(111) slab with 12 layers including vacancies on top layers. To make the numerical calculations tractable, the vacancies are assumed to distribute with the $\sqrt{3}\times\sqrt{3}$ structure as shown in Fig. 4. For the ion core pseudopotential, the analytical expression, eq.(2.7) is used with the parameters $a_1 = -0.7486$, $a_2 = 0.0768$, $a_3 = -0.2141$ and $a_4 = 0.4097$ in a.u.. The iteration procedures for the valence wave function, the valence charge density, and the crystal potential are repeated as described in section 2, until the selfconsistency is attained.

Fig. 5 shows the dispersion of the dangling bond surface states for the slab without vacancy in the surface Brillouin zone of the $\sqrt{3}\times\sqrt{3}$ structure. There are three pairs of the dangling bond bands with the Si $3p_z$ character corresponding to three atoms on each extended

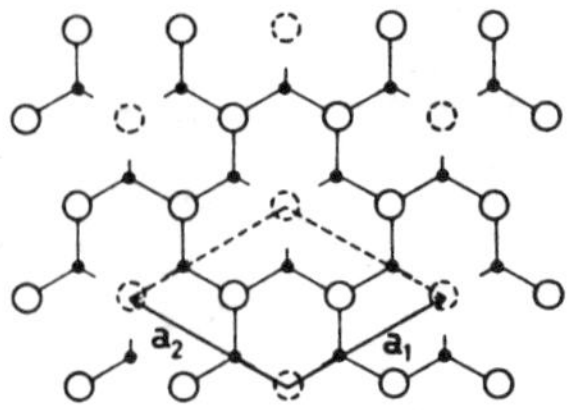

FIG. 4. Vacancy model of Si(111) $\sqrt{3}\times\sqrt{3}$ structure

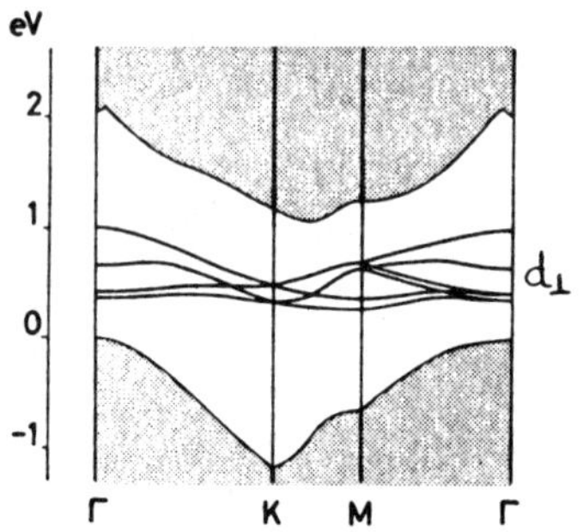

FIG. 5. Dispersion of the dangling bond for the ideal Si(111) surface in the Brillouin Zone of $\sqrt{3}\times\sqrt{3}$ structure.

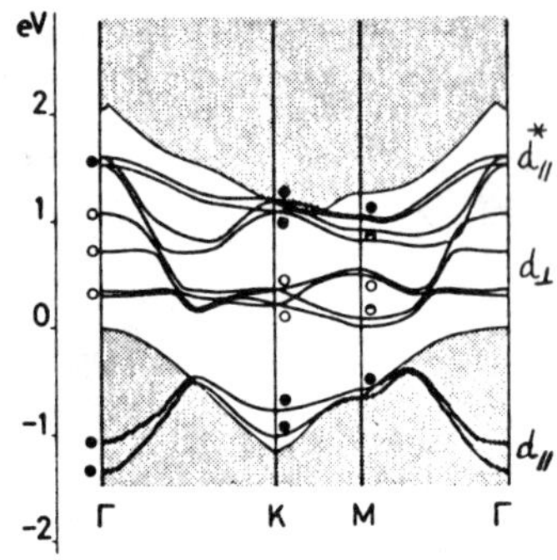

FIG. 6. Dispersion of the dangling bonds for the Si(111) $\sqrt{3}\times\sqrt{3}$ surface with vacancy.

surface unit cell. The splitting between paired bands are found to be less than 0.3 eV, which is caused by the interaction of the surface states on the two surfaces. The dispersion of the surface states for the slab including vacancies in the $\sqrt{3}\times\sqrt{3}$ structure is shown in Fig.6. There are two pairs of bands of the Si $3p_z$ ($z_\perp$ surface) like dangling bonds (hereafter called as $d_\perp$), since there are two top Si atoms on each surface layer. Each surface includes one vacancy per surface unit cell. Thus there are three pairs of energy bands corresponding to the dangling bond ($d_{\parallel}$, $d_{\parallel}^*$) directed towards the vacancy from the nearest neighboring Si atoms. Due to the interaction among them, these dangling bond states are split into two groups; one with the lower energy ($d_{\parallel}$) corresponds to the bonding state, and the other with the higher energy ($d_{\parallel}^{*}$) to the antibonding states. Because of the contributions from both surfaces, the $d_{\parallel}$ states consist of two bands, while the $d_{\parallel}^{*}$ states of four bands.

The local state densities of the first, second and the third double layer in the slab are shown respectively in Fig. 7. The sharp three peaks of $d_{\perp}$, $d_{\parallel}$, $d_{\parallel}^{*}$ are clearly seen in the top double layer. These peaks decay rapidly in the deeper regions, and the local state density at the 5-th through 6-th layer is almost the same as the state density in the bulk, which is shown by the dashed curve in Fig. 7. The peak of the $d_{\perp}$ state appears at the energy close to the Fermi level, E_F. On the other hand the peak of $d_{\parallel}$ appears about 0.9 eV below the $d_{\perp}$ state. The appearance of the two types of the dangling bond states, $d_{\perp}$ and $d_{\parallel}$ is inherent in the vacancy model and does not depend on the arrangement of the vacancies.

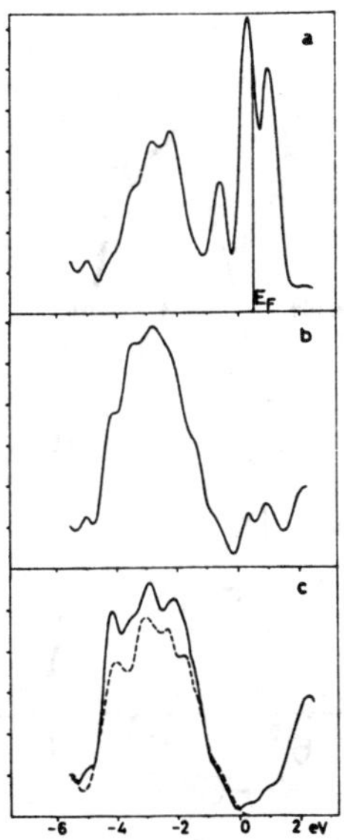

FIG. 7. The local state densities for the Si(111) $\sqrt{3}\times\sqrt{3}$ surface with vacancy. a, b, c correspond to the top double layer, the second double layer and the third double layer, respectively.

Guicher et al[31)] observed two photoelectron peaks of the surface states at 0.9 eV and 1.8 eV below E_F with the photon energy $h\nu$ = 52 eV (see Fig. 8). The peak with the smaller binding energy is enhanced by the p-polarized light, while the one with the larger binding energy becomes intensified by the s-polarized light. Therefore the shallower and the deeper peak are predominantly of Si $3p_z$ character and Si $3p_{x,y}$ character, respectively. One can assign that the shallower peak corresponds to the $d_\perp$ dangling bond and the deeper peak to the $d_\parallel$ dangling bond.

The experimentally observed separation between the two peaks is about 0.9 eV, consistent with the theoretical peak energy difference between $d_\perp$ and $d_\parallel$ band. However, the agreement of the peak energy difference cannot be taken so seriously, because the value changes sensitively with the relaxation[8,9] and the distribution of vacancies. The peak energy of $d_\perp$ is observed at somewhat deeper position than E_F in the UPS experiments. This effect would be explained by the many-electron process which is important for the photoemission from a narrow band.

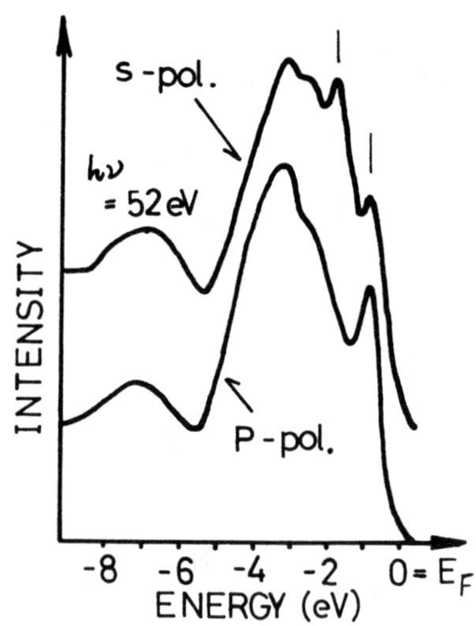

FIG. 8. Normal photoemission spectra of Si(111) 7×7 surface by Guichar et al.

3.4 Electronic structure of cluster models of Si(111) surface

We investigated the electronic structure of chemisorption systems on Si(111) surface and clarified the effect of vacancies.[10-12] For the numerical calculations of the electronic structure, the DV-Xα cluster method was used. The validity of the cluster model was

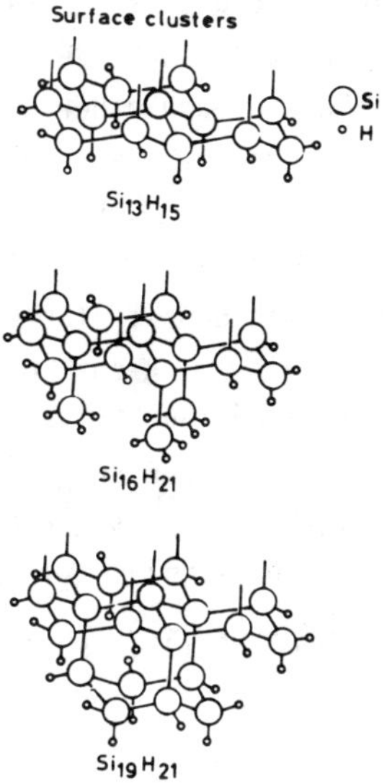

FIG. 9. Structure of model clusters, $Si_{13}H_{15}$, $Si_{16}H_{21}$ and $Si_{19}H_{21}$ for the Si(111) surface.

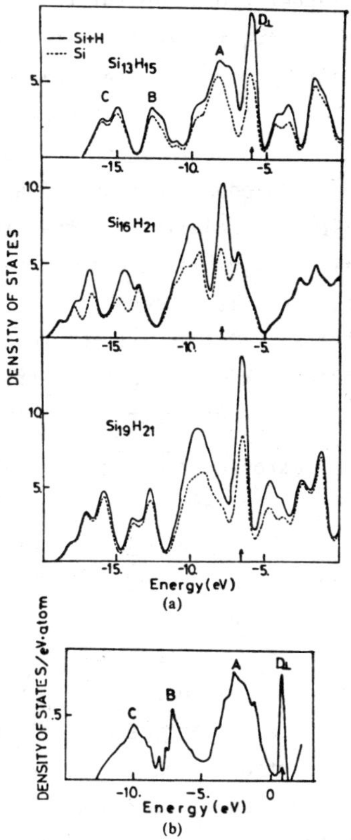

FIG. 10. a) State densities for the clusters, $Si_{13}H_{15}$, $Si_{16}H_{21}$ and $Si_{19}H_{21}$. The dashed curve shows the partial state density without hydrogen orbital components. b) The state density of the 12-layer slab for the ideal (111) surface by Schlüter et al.[6]

examined with the clusters $Si_{13}H_{15}$, $Si_{16}H_{21}$ and $Si_{19}H_{21}$, representing the Si(111) surface[11,12] (Fig. 9). The hydrogen atoms are attached to the peripheral Si atoms to saturate their dangling bonds. The Si-H bond length is assumed as 2.80 a.u. Figure 10 shows the state densities of the three clusters and compares them with the result by the selfconsistent pseudopotential method for the slab.[6] The overall features of the state densities for $Si_{13}H_{15}$ and $Si_{19}H_{21}$ are similar to that by slab, although the peak of the d state is enhanced because of the larger proportion of the top Si atoms in the clusters. For $Si_{13}H_{15}$ and $Si_{19}H_{21}$, the total state density and the partial state density without hydrogen orbital components are similar to each other. Thus the bond saturation by the hydrogen atoms does not cause any extra peaks and the Si-H bond is considered to resemble very much the Si-Si bond. Hydrogen atoms were found to be almost neutral by the Mulliken population analysis. On the other hand, the state density for the cluster $Si_{16}H_{21}$ is distorted remarkably as compared with that of the slab. This is caused by the inclusion of the incomplete "six-membered-rings" in this cluster, whereas the clusters $Si_{13}H_{15}$ and $Si_{19}H_{21}$ do not contain such an inadequate structure. Therefore we used the cluster $Si_{13}H_{15}$ as the substrate for the chemisorption on Si(111) surface.

3.4 Electronic structure of chemisorption systems, Si(111)/H,Cl

The total and the partial state densities of the cluster, a) $Si_{13}H_{15} + H_7$ and b) $Si_{12}H_{15} + H_6$ are shown in Fig. 11a, b, respectively.[11,12] The hydrogen atoms are chemisorbed on top of the surface Si atoms with the Si-H distance 2.8 a.u. (Fig. 12). In the cluster $Si_{12}H_{15} + H_6$, the central surface Si atom (Si(1)) is missing and no H atoms are chemisorbed on the vacancy site. For both clusters, it is clearly seen that the peak of the d state is removed by the hydrogen chemisorption. The chemisorption induces a strong peak A at -10.5 eV and a subsiduary peak B at -12.5 eV. These characteristics agree well with the UPS result obtained for Si(111) 7×7/H surface.[32] Although the overall features of the electronic structure induced by the chemisorption are same for both the clusters, the following difference is noticed between them. Namely in the valley region (∿ -11.5 eV) between the two chemisorption induced structures, the state density increases for the cluster with vacancy, while it decreases for the cluster without vacancy. The former but not the latter characeristics are observed by the UPS experiments.[32]

Figure 13a illustrates the total and the partial state densities of the cluster c) $Si_{12}H_{15} + Cl_6(I) + Cl(III)$ modeled on the chlorine chemisorbed surface with vacancy.[10-12] On the other hand, Fig. 13b shows the state densities of the cluster, d) $Si_{13}H_{15} + Cl_6(I) + Cl(I')$ for the Cl chemisorbed on buckled surface.[10-12] For both the clusters, six Cl(I) atoms are assumed to chemisorbe on top of the Si(2)

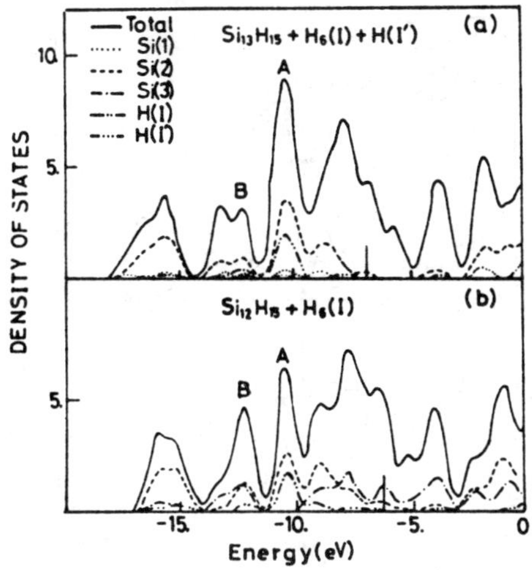

FIG. 11. The total and the partial state densities of a) the cluster $Si_{13}H_{15} + H_7$ and b) $Si_{12}H_{15} + H_6$. Vertical line indicates the Fermi level.

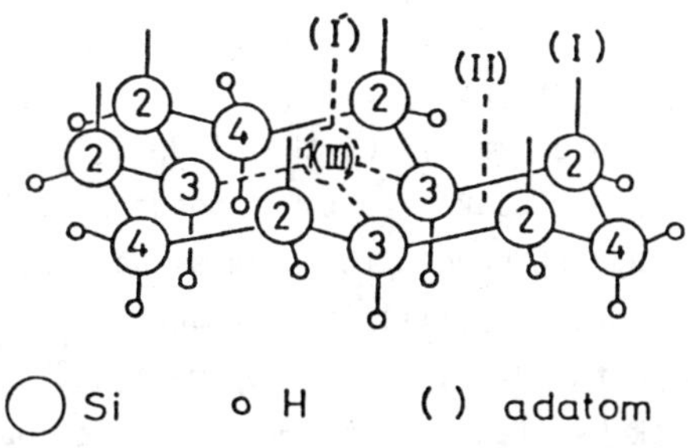

FIG. 12. The structure of the cluster $Si_{13,12} + A_{3,6,7}$ where A indicates the adatom.

atoms with the Cl(I)-Si(2) distance 3.8 a.u. (see Fig. 12). In the cluster c), the Cl(III) atom is also chemisorbed at 2.0 a.u. directly above the vacancy site. In the cluster d), the central Si(1) atom is raised and the six Si(2) atoms are lowered by 0.5 a.u., respectively, from the original position of the ideal surface. The Cl(I')-Si(1) bond length is assumed to be 3.8 a.u., just the same as Cl(I)-Si(2) bond length.

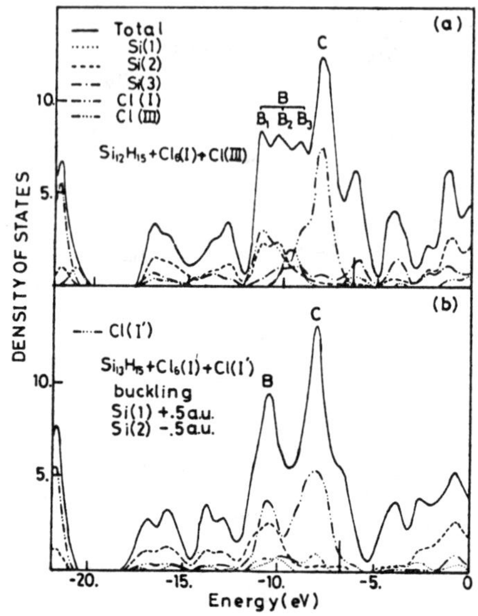

FIG. 13. The total and the partial state densities of a) $Si_{12}H_{15}+Cl_7$ and b) $Si_{13}H_{15}+Cl_7$.

For the case of the buckled substrate, two sharp peaks are induced by the chlorine chemisorption. The higher peak C corresponds to the π bonding states contributed mainly by the Cl $3p_x$, $3p_y$ orbitals (x, y surface), while the lower peak B to the σ bonding state made of the Cl $3p_z$ and the $d_\perp$ dangling bond. The peak structure explains well the UPS spectra of Cl adsorbed Si(111) 2×1 surface by Rowe et al.[33] They observed two peaks in the UPS with the separation about 2.1 eV. For the s polarization (E // surface), the lower peak was observed to be greatly suppressed indicating that this is mainly made of the Cl $3p_z$ orbital.

On the other hand the peak B for the cluster, $Si_{12}H_{15}+Cl_6(I)+Cl(III)$ has a complicated structure as shown in Fig. 13a. This is because the σ bonding states of Cl(III)-Si(3) have the energy close to that of the Cl(I)-Si(2) σ bonding states and contributes to the peak B. One expects, therefore, that the peak B of the Cl covered Si(111) surface is not suppressed so much for the s polarized light, if the substrate includes vacancies. In fact, such characteristics are observed in the UPS of the Cl covered Si(111) 7×7 surface.[33)]

As the conclusion of this section, the calculated electronic structure of Si(111) surface with vacancy explains fairly well the observed UPS spectra on the 7×7 reconstracted surface, especially the photon polarization dependence of the surface states. The higher photoemission peak is assigned to the vertical dangling bond $d_{\perp}$ and the lower peak to the lateral dangling bond $d_{//}$ around the vacancy. The photoemission peaks induced by the chemisorptive bonds of H and Cl are also well explained by the electronic structure calculated for the proper cluster models. Some features in UPS for chemisorptions on Si(111) 7×7 surface seem better explained by the model with the vacancy than the model without it. Thus the electronic structure obtained by theoretical calculations seems to support the vacancy model of the Si(111) 7×7 surface. Of course much more complementary studies should be done to establish the model unambiguously.

4. Electronic Structure of the Polar Surfaces

4.1 How the charge neutrality is attained?

Compared with the metals or elemental semiconductors, the surface electronic structure of compounds is much more complicated. This is due to the fact that the charges of surface ions can change from those of the bulk. It seems the rule that the covalency of the ionic compound is enhanced and the ionicity is reduced on the neutral surface.[13)] The most drastic change is found on the polar surfaces of compounds.[14,15)] If the charges of the surface ions were the same as those of the bulk ions, the structure of the polar surface would be unstable. This is due to the macroscopic field in the bulk produced by the periodic array of the double layers (see Fig. 14) The macroscopic field should be cancelled out by somehow induced surface charge. The area density of the induced charge σ_c should be

$$\sigma_c = \sigma_b d/a \tag{4.1}$$

where σ_b is the area charge density of the bulk layer, d and a are the thickness of the double layer and the displacement between the adjacent double layers, respectively.[15)]

In many polar surfaces, this compensating charge is introduced by surface reconstructions such as vacancies or faceting.[22)] For example,

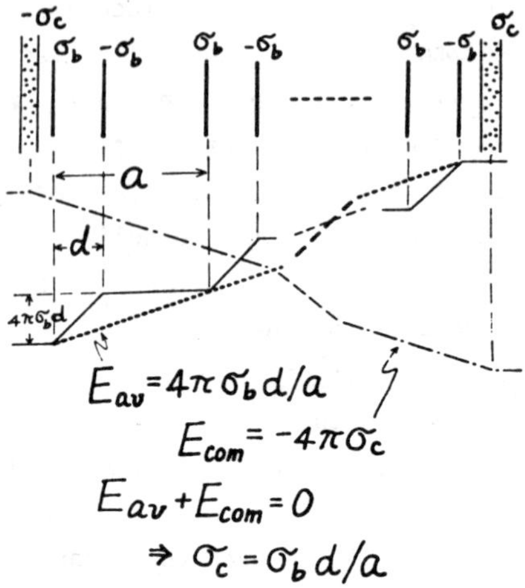

FIG. 14. The cancellation of the macroscopic field E_{av} by the compensating charge.

vacancies on the cation (anion) surface induce effective negative (positive) charge which cancels the bulk macroscopic field. The facet structure on the polar surface also generates the compensating charge as discussed in ref.22. Most reconstruction structures on the polar semiconductor surfaces are explained by such mechanisms.

The electronic structure of unreconstructed polar surfaces is quite interesting, because the charge state must be changed drastically as compared to the bulk to introduce the compensating charge. Such peculiar electronic structure might reflect on the properties of the surface, and would cause its special chemical activity.

4.2 Electronic structures of polar surfaces of ZnO

It is known that the unreconstructed polar ZnO surface can be easily prepared.[2,34] Surface states and chemisorptions on the surface have been experimentally studied.[2,34] The electronic structure of the ZnO(0001)Zn, and ZnO(000$\bar{1}$)O surfaces have been studied by the DV-Xα cluster calculations.[14] Two model clusters, $Zn_{10}O_{10}B$ and $Zn_{10}O_{10}S$ were used to represent the polar surfaces. They are shown in Fig. 15b, c, respectively. The total state densities for these clusters are shown in Fig. 16 together with that of the cluster Zn_4O_{13} (see Fig. 15a) without Zn surface. The overall state density of the three clusters in the oxygen valence band and Zn 3d band resembles each other and explains fairly well the observed UPS.[35] However, although the band gap between the conduction and the valence band clearly exists in the Zn_4O_{13} cluster, it disappears for the clusters with Zn polar surfaces. As shown in the partial state densities of the $Zn_{10}O_{10}S$ cluster depicted in Fig. 17, the states in the gap region are mainly contributed by the 4s, 4p states of the surface Zn ions. These gap states are the dangling bonds generated by the removal of the nearest neighbor O ions. It should be remarked that the more reduced the Zn coordination, the more enhanced the admixture of its back bonds into the O 2p band. For the extreme case of the corner Zn(2) ion in the $Zn_{10}O_{10}S$ cluster, the local state density shows simple metallic character as a whole.

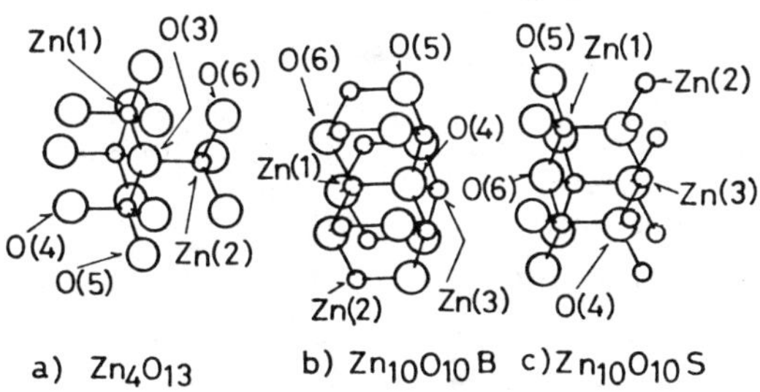

FIG. 15. Structure of the clusters a) Zn_4O_{13}, b) $Zn_{10}O_{10}B$, and c) $Zn_{10}O_{10}S$.

The charge distributions of the clusters obtained by Mulliken population analysis are listed in Table I. The charges of the bulklike ions, Zn(1), O(4) are the largest (1.3 for $Zn_{10}O_{10}B$ and 1.2 for $Zn_{10}O_{10}S$) and roughly equal to the value inferred by experiments. Compared to this, the charges of the surface ions are considerably reduced. The charges on the corner ions (Zn(2), O(5)) are much more reduced than those of the bulklike and the surface ions.

If one defines the bulk charge σ_b by

$$\sigma_b \sim 0.5 \times \{|\text{inner cation charge}| + |\text{inner anion charge}|\} \tag{4.2}$$

and the surface charge σ_s by

$$\sigma_s \sim 0.5 \times \{|\text{surface cation charge}| + |\text{surface anion charge}|\}, \tag{4.3}$$

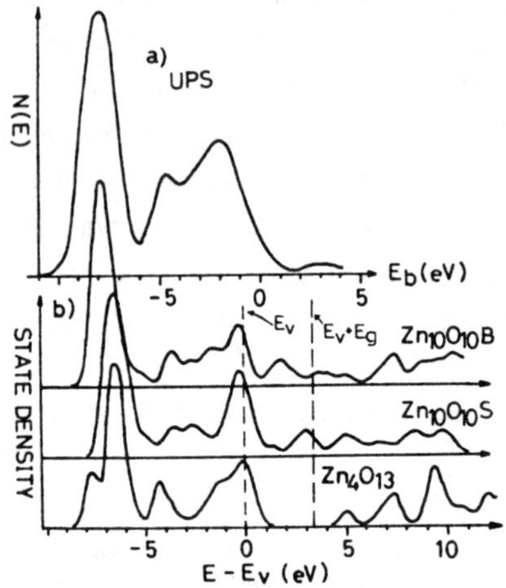

FIG. 16. a) Photoemission spectrum by Ranke[35] ($h\omega = 48.7$ eV) for ZnO(0001)Zn surface and b) the state density of Zn_4O_{13}, $Zn_{10}O_{10}B$ and $Zn_{10}O_{10}S$.

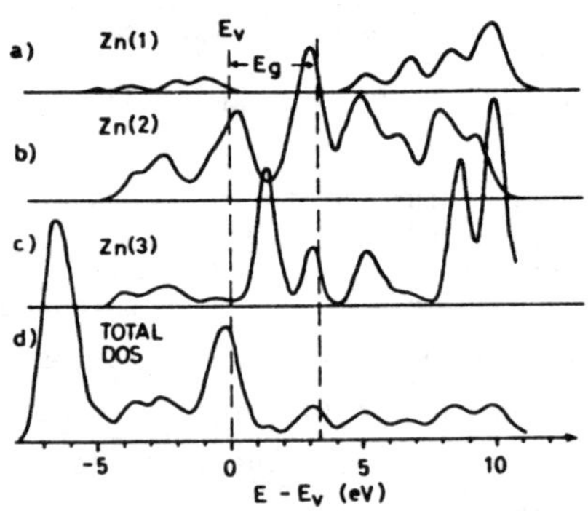

FIG. 17. Partial state densities for the 4s, 4p components of a) Zn(1), b) Zn(2) and c) Zn(3) ion in $Zn_{10}O_{10}S$ cluster. The total state density is also shown in d).

the surface charge relative to the bulk σ_s/σ_b is obtained as shown in Table II. For the surface ion, only Zn(3) and O(6) are used excluding the corner ions. The ratio of the compensating charge σ_c to the bulk charge

$$\sigma_c/\sigma_b = 1 - \sigma_s/\sigma_b \tag{4.4}$$

Table I. Charge of ions in several clusters.

Custers	bulklike ions		Corner ions		surface ions		adatom
	Zn(1)	O(4)	Zn(2)	O(5)	Zn(3)	O(6)	O(7)
$Zn_{10}O_{10}B$	1.33	-1.31	0.83	-0.87	1.07	-1.00	
$Zn_{10}O_{10}S$	1.18	-1.16	0.38	-0.42	0.97	-0.83	
$O\text{-}Zn_{10}O_{10}S$	1.17	-1.14	0.45	-0.40	0.98	-0.84	-0.53

Table II. Compensating charges of polar surfaces obtained by the cluster method.

Custers (Surfaces)	$Mg_{13}O_{13}$ (MgO(111))	$Zn_{10}O_{10}B$ (ZnO(0001))	$Zn_{10}O_{10}S$ (ZnO(0001))
σ_s/σ_b	0.49	0.78	0.77
σ_c/σ_b	0.51	0.22	0.23
d/a	0.50	0.25	0.25

is also shown in Table II. The value is very close to theoretical one, i.e. d/a for the infinite surface.[14,15] The reduction of the positive charge on the surface and the corner Zn ions are partly due to the crossing of the dangling bond band with the Fermi level, E_F. The greatly enhanced covalent mixing of the back bonds due to the reduction of the cation-anion energy difference is the other mechanism of the charge reduction.

That the states locating in the gap region are the dangling bonds is clearly seen by observing the change of the electronic states with the oxygen chemisorption onto the Zn surface.[14] Figure 18 illustrates the total and the local state densities of the cluster $O\text{-}Zn_{10}O_{10}S$ (Fig. 19), which is obtained by the chemisorption of oxygen atom (O(7)) onto the central Zn ion (Zn(3)) of the cation surface of $Zn_{10}O_{10}S$. As shown in Fig. 18, the sharp peak of Zn(3) in the gap region (see Fig. 17c) disappears with the oxygen chemisorption. This indicates that the strong chemisorption bond is formed by the interaction of O(3) $2p_z$ orbital and Zn(3) $4s+4p_z$ type dangling bond. The energy of the chemisorption bonds appears at about 2 eV and 4 eV below the valence band top.

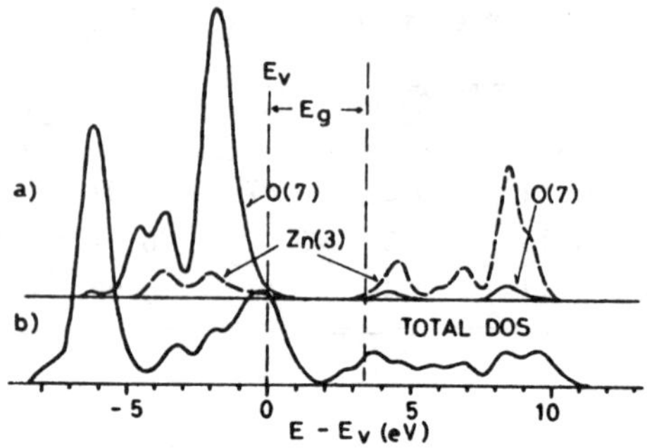

FIG. 18. a) The local state densities at O(7) (full line) and Zn(3) (dashed line) in the cluster O-$Zn_{10}O_{10}S$. b) The total state density of O-$Zn_{10}O_{10}S$ cluster.

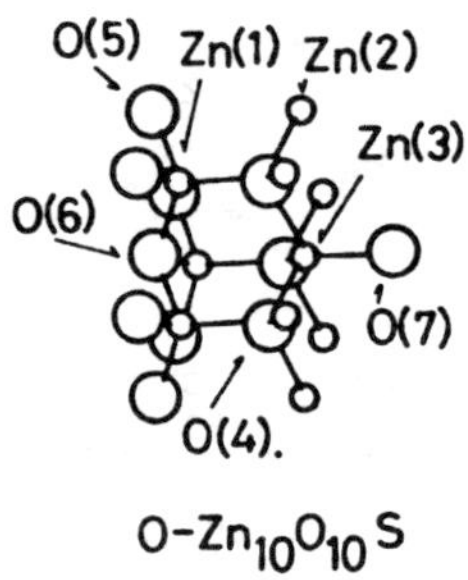

FIG. 19. Structure of the cluster O-$Zn_{10}O_{10}S$.

4.3 Electronic structure of MgO(111) and MgO($\bar{1}\bar{1}\bar{1}$) surface

Although MgO(111) polar surface is less stable than the (100) neutral surface, it can be prepared by several methods. It was found by Miyahara et al[36] that the MgO(111) surface reveals catalytic activity in butadiene hydrogenation reaction, while the ideal (100) surface has not such an activity. As discussed below, the peculiar electronic structures of the cation polar surface seems responsible for it.

DV-Xα calculation has been performed[15] for the cluster $Mg_{13}O_{13}$ including the (111)Mg and ($\bar{1}\bar{1}\bar{1}$)O polar surfaces of the MgO crystal (Fig. 20). The local state densities of the respective ions are shown in Fig. 21. The electronic structure of inner ions, Mg(5) and O(3) are almost the same as that expected by the bulk energy band, except a small structure B reflecting the surface state tail above the valence band A. The local state densities at Mg ions on the cation surface

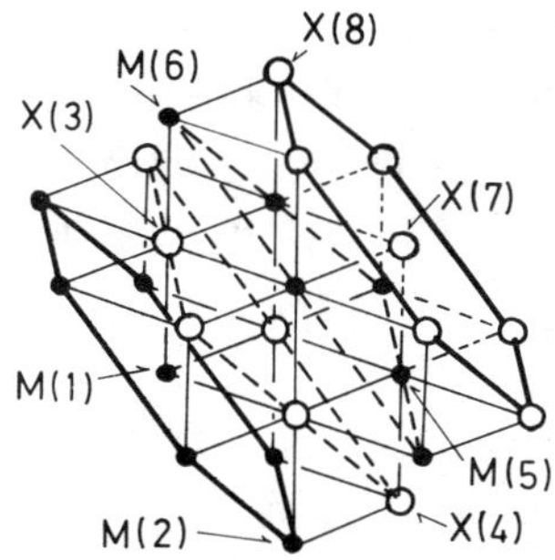

FIG. 20. Model cluster $M_{13}X_{13}$ for the polar MX(111) surface of the rock-salt crystal.

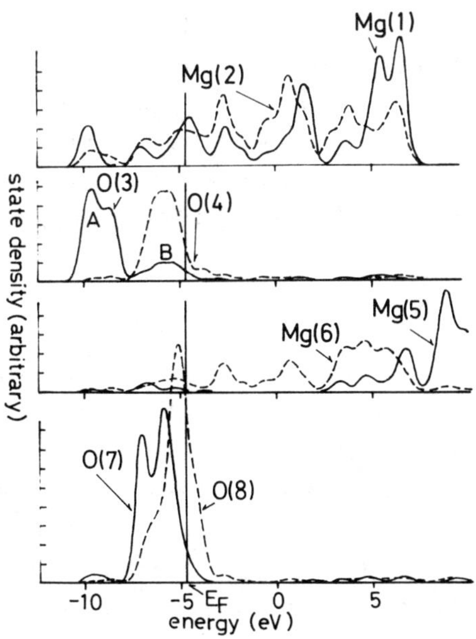

FIG. 21. The local state density at each ion in $Mg_{13}O_{13}$ cluster.

(Mg(1), Mg(2)) reveal quite different structure. The state density on this surface is similar to that of simple metal and retains no traces of the band gap. Since the continuous energy band extends below E_F, electrons flow into the top Mg layer. On the other hand, the O 2p bands on the anion surface are remarkably raised due to the reduction of the attractive Madelung potential. Thus the upper part of the band goes above E_F, leading some electrons to flow out of the anion layer. As the result, the compensating charge of about 50% of the bulk charge is induced on the MgO polar surfaces as listed in Table II. This value is consistent with the theoretical one,

$$\sigma_c/\sigma_b = d/a = 0.5$$

for the (111) surface of the rock-salt type crystal.

In concluding this section, it was revealed that the charges are induced on unreconstructed polar surfaces of compounds which cancel the macroscopic field in the bulk. The negative surface charge flows into the cation layer by the lowering of some cation surface states below E_F. On the other hand, positive surface charge is generated on the anion layer due to the raising of the upper surface anion band above E_F. The enhanced orbital admixtures of the back bonds is another mechanism to induce the surface charge. The electronic structure of the cation polar surface is quite different from that in the bulk. It has metallic character with no trace of the band gap. The enhanced chemical reactivity might be resulted from such special electronic states.[36)]

5. Summary

As exemplified in previous sections, the non-empirical theory based on the local electron density can reveal the detailed electronic structure of complicated surfaces. Frequently used approaches, i.e. the local density functional method, the DV-Xα method and self-consistent pseudo-potential method have been discussed in some details. From the theoretical electronic structure of surfaces, one can obtain the energy and spatial distribution of electrons, which must be consistent with experiments such as UPS, ELS, INS and the work function measurements. Therefore proper surface models can be selected among those proposed by experiments on the surface atomic structure. Moreover the theoretical picture of the electronic structure is indispensable to understand molecular processes on surfaces.

The correlation between the atomic and the electronic structure is particularly important for semiconductor surfaces. The energy and the orbital character of the dangling bond changes sensitively with the surface atomic arrangement. We have discussed that some simple reconstructed structures can be understood by the stabilization of the dangling bond energy through the rebonding mechanism or the increase of the s component due to the sharpening of the back bond angle.

The atomic arrangement and the mechanism of the Si(111) 7×7 surface are very puzzling. The electronic structure of Ino's vacancy model, which is promising from the RHEED and LEED analysis and consistent with the 7 × 7 ↔ 1 × 1 phase transition process,[30] are clarified in detail. The appearance of the two dangling bond, $d_{\perp}$ and $d_{//}$, explains fairly well the characteristics of the UPS. The vacancy model can also explains special features of the UPS observed for the chemisorption systems, Si(111)/H,Cl.

The charge state on the surfaces of compound crystals is a typical example, where the self-consistent first-principles theory provides crucial informations. It was shown that the compensating charge is generated on the surface, which is required to cancel the macroscopic field in the bulk. The mechanism of the surface charge generation is the electron transfers between the surface states and the bulk, as well as the enhanced orbital admixture on the back bonds. The enhanced chemical activity is inferred from the peculiar electronic structure of the polar surface. The drastically changed charge state on the top ions could be verified by an accurate measurement of the core level shift.

Acknowledgments

This article is mostly based on the series of works performed in co-operation with Dr. T. Hoshino and H. Nagayoshi. The author is deeply indebted to them.

References

1) Interaction on Metal Surfaces, R. Gomer ed., Springer-Verlag (1975); Photoemission and the Electronic Properties of Surfaces, B. Feuerbacher, B. Fitton and R.F. Willis, eds., Wiley (1978); Surface Physics of Materials, J.M. Blakely, ed., Academic Press (1975); J.A. Appelbaum and D.R. Hamann, Rev. Mod. Phys., 48, 479 (1976).
2) Theory of Chemisorption, J.R. Smith, ed., Springer-Verlag (1980), Berlin; S.R. Morrison, The Chemical Physics of Surfaces, Plenum, New York (1977).
3) W.A. Goddard III, J.J. Barton, A. Redondo and T.C. McGill, J. Vac. Sci. & Technol., 15, 1274 (1978).

4) P. Hohenberg and W. Kohn, Phys. Rev., 136, B864 (1964); W. Kohn and L.J. Sham, Phys. Rev. 140, A1133 (1965).
5) J.C. Slater, Quantum Theory of Molecules and Solids, vol.4, The self-consistent Field for Molecules and Solids, McGraw-Hill (1974).
6) M. Schlüter, J.R. Chelikowski, S.G. Louie and M.L. Cohen, Phys. Rev., B12, 4200 (1975).
7) A. Zunger and M.L. Cohen, Phys. Rev., B19, 568 (1979).
8) H. Nagayoshi and M. Tsukada, Surface Sci., 116, 163 (1982).
9) K. Nakamura, T. Hoshino, M. Tsukada, S. Ohnishi and S. Sugano, J. Phys. C, 14, 2165 (1981).
10) T. Hoshino and M. Tsukada, Solid State Commun., 38, 231 (1981).
11) M. Tsukada and T. Hoshino, Int. J. Quantum Chem. Symp., 15, 445 (1981).
12) T. Hoshino and M. Tsukada, Surface Sci., 115, 104 (1982).
13) M. Tsukada, H. Adachi, C. Satoko, to be appeared in Prog. in Surface Sci.
14) M. Tsukada, E. Miyazaki and H. Adachi, J. Phys. Soc. Jpn., 50, 3032 (1981).
15) M. Tsukada, T. Hoshino, J. Phys. Soc. Jpn., 8, 2562 (1982).
16) H. Adachi, M. Tsukada and C. Satoko, J. Phys. Soc. Jpn., 45, 874 (1978).
17) N.D. Lang, in Solid State Phys. 28, H. Ehrenreich, F. Seitz and D. Turnbull eds., p.225, Academic Press (1973).
18) D. Pines, Elementary Excitations in Solids, Benjamin (1963).
19) D. Pines and P. Nozierès, The Theory of Quantum Liquids, Benjamin (1966).
20) J.A. Appelbaum and D.R. Hamann, Phys. Rev., B8, 1777 (1973).
21) J. Chadi, Proc. Int. Conf. Physics of Semiconductors, Kyoto, 1980, J. Phys. Soc. Jpn., 49, Suppl.A, P.1035 (1980).
22) R.W. Nosker, P. Mark and J.D. Levine, Surface. Sci., 19, 291 (1970).
23) K. Suzuki and T. Hoshino, Proc. Int. Conf. Physics of Semiconductors, Kyoto, 1980, J. Phys. Soc. Jpn., 49, Suppl.A, p.1055 (1980).
24) S.G. Louie, J.R. Chelikowski and M.L. Cohen, J. Vac. Sci. Technol., 13, 790 (1976).
25) J.J. Lauder and J. Morrison, J. Appl. Phys., 34, 1403 (1963).
26) S. Ino, Jpn. J. Appl. Phys., 19, L61 (1980), 19, 1277 (1980).
27) D.J. Chadi, Surface Sci., 99, 1 (1980).
28) K.C. Pandey, to be published.
29) A. Harrison, Surface Sci., 55, 1 (1976).
30) N. Osakabe, K. Yagi and G. Honjo, Jpn. J. Appl. Phys., 19, L309 (1980).
31) G.M. Guicher, F. Houzay, R. Pinchaux and Y. Petroff, Proc. Int. Conf. Physics of Semiconductors, Kyoto, 1980, J. Phys. Soc. Jpn., 49, Suppl.A, p.1047 (1980); F. Houzay, G.M. Guicher, R. Pinchaux and Y. Petroff, J. Vac. Sci. Technol., 18, 860 (1981).
32) T. Sakurai and H.D. Hagstrum, Phys. Rev., B12, 5349 (1975).
33) J.E. Rowe, G. Margaritondo and S.B. Christman, Phys. Rev., B16, 1581 (1977).
34) C.G. Scott and C.E. Reed, in Surface Physics of Phosphors and Semiconductors, C.G. Scott and C.E. Reed eds., p.411, Academic Press (1975).
35) W. Ranke, Solid State Commun., 19, 685 (1976).
36) K. Miyahara, Y. Murata, I. Toyoshima, Y. Tanaka and T. Yokoyama, J. Catal., 68, 186 (1981).

SOME RECENT ADVANCES IN BASIC VLSI TECHNOLOGY

Yoshiyuki Takeishi

Toshiba Research and Development Center
Toshiba Corporation
Komukai, Saiwaiku, Kawasaki 210, Japan

ABSTRACT

Among recently developed technologies for very-large-scale integrated circuits (VLSI), some of the topical problems that are closely associated with Si material science and device physics are reviewed. Selected issues are: 1) new crystal growth technology, 2) effects of grown-in oxygen impurity in the starting material and microdefect formation upon high-temperature heat treatment on device fabrication, 3) solid-phase regrowth epitaxy of very thin Si films on sapphire, 4) structural design constraints in miniaturizing MOS transistors and 5) effects of non-thermal hot-carrier generation on device reliability.

1. Introduction

As is widely known, the number of integrated active devices on a silicon chip in commercial IC products has been increased at a rate of almost twice a year over twenty years since 1960. This is a result of combined improvements of a variety of technologies concerning device structure, circuit design, crystal growth, defects control, pattern lithography and wafer processing such as oxidation, impurity diffusion or ion implantation, insulator deposition and metallization. Table I summarizes a progress in the structural dimensions of the MOS type dynamic random access, read-write memory(d-RAM), which is the most popular, but the most sophisticated LSI, representative of the front of cost-effective integration technology in every corresponding period.

When going from one period to another, some basic problems creeping in the past products appear to be solved in parallel to production technology developments. In this brief article, we cannot cover all aspects of recent progress in VLSI technology. Instead, we limit ourselves to focus on some of the recent achievements in studying defects behaviors in Si wafers and design constraints in miniaturizing MOS transistors, which will be deliberately utilized in advanced VLSI products in a few years to come.

Table I. Technology of MOS dynamic Random-Access Memory.

Number of bits (k = 1024)	1k	4k	16k	64k	256k	1024k*
Product release	1970	1973	1976	1980	1983	1986
Gate structure/material	single poly Si	single poly Si	double poly Si	double poly Si	refractory metal silicide	refractory metal
Channel type	p	n	n	n	n	n, p
Number of devices/memory cell	3	3	2	2	2	2
Cell area (μm^2)	3,000	1,200	500	150	75	30
Gate oxide thickness (Å)	1,200	1,000	800	450	250	150
Effective channel length (μm)	10	7.5	5	2.5	1.5	1.0
Junction depth (μm)	2.5	2	1.2	0.5	0.35	0.25
Poly Si line (μm)/space (μm)	10/10	7.5/7.5	5/6	3/4	1.5/1.5	1.2/1.2
Al line (μm)/space (μm)	10/10	7.5/7.5	5/6	3/4	2/1.5	1.2/1.2
Electrode contact opening (μm)	10 x 10	7.5/7.5	4 x 4	2.5 x 2.5	1.5 x 1.5	1.2 x 1.2
Chip area (mm^2)	11	16	20	25	45	60
Total active devices/chip	4×10^3	1×10^4	3.7×10^4	1.5×10^5	6×10^5	2.2×10^6
Power supply (V)	-16, 3	12, 5, -5	12, 5, -5	5	5	3

* predicted

2. Defects Control of Silicon Wafers

2.1 Crystal Growth

Figure 1 is a photograph of a world-largest, Czochralski(CZ) silicon ingot as large as 200 mm in diameter, which is perfect so much so that an X-ray topograph measurement instrument cannot detect any dislocations. Although the size is far beyond the currently used wafer sizes of 100 or 125 mm, it manifests itself in the great progress of crystal growth technology. Note that in 1970 the wafer size in IC factory was only 50 mm.

Recently two new methods were developed to improve the CZ method to grow the ingots of better quality. One is the use of pure and dense silicon-nitride coated crucible instead of conventional silica crucible, which successfully leads to a remarkable decrease in oxygen concentration (less than 2×10^{16} atoms/cc, detection limit).[1] Additional use of silicon-carbide coated carbon heater further decreased the carbon concentration to less than 5×10^{15} atoms/cc, detection limit.* Resistivity of the new CZ crystals can be increased to as high as 1000 Ωcm. Intentional dopings of oxygen and carbon are readily done, whenever it is necessary.

Another new technology is the application of transverse magnetic field during the CZ growing process,[2] which prevents thermal fluctuation and thermal convection in molten state. This method yields the crystal of uniform radial distribution of low oxygen concentration (less than 1×10^{17} atoms/cc) and less point and cluster defects. The resistivity is 400 Ωcm. This is called MCZ method. More recently, the application of vertical magnetic field was also successfully studied to grow CZ crystals of uniform axial distribution of impurities.[3]

2.2 Wafer Bow and Warpage

When one looks closely at wafers carelessly sliced from ingots, and wafers subjected to careless high-temperature heat cycles, these wafers are rarely flat enough, but are geometrically deformed to be spherical, cylindrical, saddle and even tengallon-hat shapes. An extensive study was recently performed to elucidate the origins of these deformations.[4]

* Toshiba Corporation and Toshiba Ceramics, private communication.

Figure 1. Photograph of a silicon ingot of 200 mm in diameter. (Courtesy of Toshiba Corporation.)

A few years ago, the magnitude of bow of sliced and polished wafers supplied by various wafer manufacturers was seemingly very complicated. Some wafers of 75 mm in diameter showed a bow of 30 μm high. A systematic study[4)] led to the conclusion that the bow has no relation to crystal growing methods, crystallographic orientation and surface polishing of either or both sides of sliced wafers, but it originates from curved slicing caused by initial slicing blade rim bending. This finding encouraged the manufacturers to be careful enough to supply flat wafers within 3 μm convex or concave.

Appreciable wafer warpage often appears upon heat cycles. This is probably due to dislocation motion. In fact, the warpage of wafers with higher dissolved oxygen concentration is smaller after heat cycles between room temperature and 1000 - 1200°C. The dissolved oxygen is thought to prevent dislocation movement. In contrast, the wafers of higher concentration of intentionally precipitated oxygen show larger warpage, suggesting that the oxygen precipitates act as a site of dislocation generation.[4] Generally, CZ wafers are mechanically stiffer than FZ (float zone) wafers that have lower oxygen concentration. Manufacturers try to strengthen FZ crystals by incorporating nitrogen. A complicated matter is that dissolved oxygen atoms likely change to oxygen precipitates around carbon atoms upon long annealing at lower temperatures in the range 650 to 800°C. Thus, the loading and unloading of the wafers to high-temperature furnaces must be very careful considering the wafer specifications. Moreover, oxidized, nitridized, or impurity-diffused wafers, which have a stress or strain, possibly make the wafer warpage complicated.

The wafer warpage which may change its shape and magnitude by successive different processings is serious for successive printings of a series of mask patterns. All series patterns should be printed on a wafer with high accuracy of less than ± 0.5 μm deviation. Advanced optical mask alignment equipments of mirror reflection types and of ten (or five)-to-one pattern-size reduction, step-and-repeat projection types for 3 to 1 μm line and space VLSI patterns tend to use shorter wave length which may mean shorter depth of focus. If vacuum-chucking mechanism cannot compensate the wafer warpage, the alignment accuracy is different on central and peripheral parts of the wafer. Otherwise, chip-by-chip local alignment instead of full wafer alignment should be necessary, which results in lower productivity. It may be easily understood how serious is a different lateral deformation of each printed pattern for next printing. When we go to submicron VLSI patterns, the problem might be very serious.

2.3 Extrinsic and Intrinsic Defects, and Impurity Gettering

Generally a tiny defect would make a finished wafer fatal. The defects are of intrinsic and extrinsic origins. The wafers under various processings in IC factory always suffer environmental pollusion by inorganic and organic dusts (even bacteria) in air, active gases for reaction, chemicals and water, and from human body and production machines. Because these external pollusions are increasingly serious for VLSI, manufacturers must be very careful of how to be free from pollusion. Since the processing wafers are difficult to be entirely free from external impurities, various practical techniques to make some of these impurities inactive are explored, some of which will be described later. On the other hand, intrinsic defects are built-in imperfections and impurities in grown crystals, and moved and segregated defects evolved in processing.

In old times, we were alert on intrinsic defects for bipolar IC concerning minority carrier life time and low-frequency noise, and on extrinsic impurities (mainly alkaline atoms) for MOS IC concerning stability of threshold voltage. In recent years, however, it turned out that even MOS LSI are sensitive to intrinsic defects as they affect, for example, holding time of charge stored in a capacitor of memory cell in d-RAM and current leakage in CMOS RAM that features extremely low power consumption in quiescent state. Concurrently LSI manufacturers found empirically that the production yield of good chips depended upon the crystal growth method and condition. For example, it was said somewhere that the yield of MOS LSI is higher by the use of CZ wafers than FZ wafers. Along this trend, extensive studies of dynamic behaviors of the grown-in defects have been done over past few years.

In general, conventional CZ crystals contain higher concentration of oxygen ($> 10^{17}$ atoms/cc) than conventional FZ crystals ($< 10^{15}$/cc). That CZ crystals contain a considerable number of oxygen atoms has other effects in addition to good stiffness. Dissolved oxygen atoms in grown CZ crystals are thought to be in a super-saturated state formed during the cooling down stage of crystal growth. The dissolved oxygen would change its state upon long high-temperature annealing. Recent studies[5)] revealed the following experimental results by isochronal annealing of CZ wafers with higher carbon concentrations ($> 2 \times 10^{16}$ atoms/cc) (Type I) and lower ones ($< 2 \times 10^{15}$ atoms/cc) (Type II): (1) at the temperatures lower than 800°C, point-like microdefects and dislocation pairs build up, (2) at 800 - 1100°C, plate-like precipitates and resultant punch-out dislocation loops appear, (3) in higher temperatures than 1100°C, large octahedral precipitates are observable and (4) these phenomena are likely to occur in Type I as shown in Fig. 2.[5)] When the wafers are heat-treated at 650°C for 64 hr, the decrease of interstitial oxygen atoms appreciably

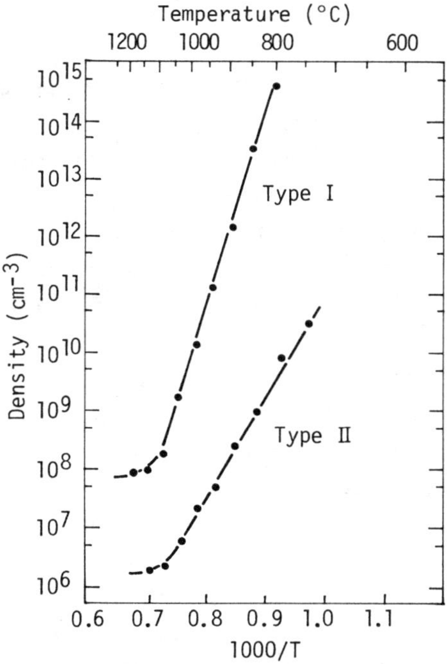

Figure 2. Oxygen precipitate density induced by isochronal annealing for 64 hr as a function of reciprocal temperature. CZ crystals: Type I is carbon rich (> 5×10^{16}/cc) and Type II is carbon lean (< 5×10^{15}/cc). After Matsushita (Ref. 5).

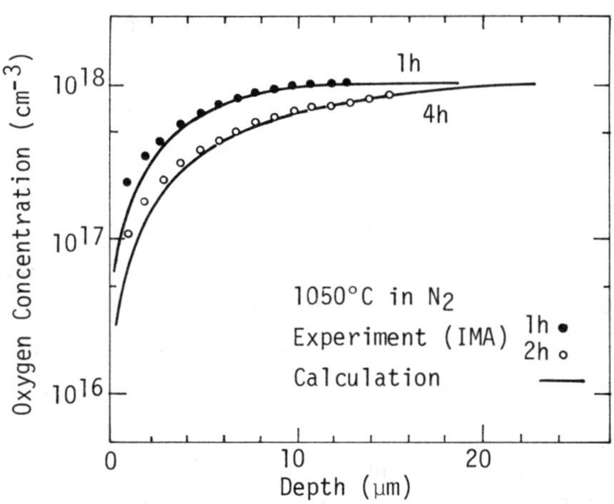

Figure 3. Oxygen out-diffusion and denuded zone formation by heat treatment at 1050°C in N_2 atmosphere. Solid curves are calculated results. After Matsushita (Ref. 5).

depends upon the carbon concentration. This fact implies that carbon or carbon-contained complex acts as an embryo of the microdefect of oxygen precipitates. The dislocation pairs, dislocation loops and stacking faults formed in the annealings at the higher temperatures are thought to result from excess interstitial silicon atoms induced by the oxygen precipitates formation.

These facts are very important for LSI. Bipolar IC, PMOS, NMOS and CMOS LSI are fabricated on different high-temperature heating programs. Including the importance of collaboration of wafer suppliers and IC manufacturers, thermal history, and oxygen and carbon concentrations must be specified in details, and thermal processes must be carefully designed for specific type of LSI.

An intentional utilization of the microdefects has been developed.[5 - 13)] The large number of microdefects can absorb metallic impurities creeping in various processings of wafers. We call this new technique as intrinsic gettering. The intrinsic gettering is of two-step processes. Under non-oxidized ambient, oxygen atoms are first out-diffused at high temperature, say, 1000°C, to form a denuded zone in the surface layer (see Fig. 3). Then, the wafers are subjected to long heating at lower temperature, say, 650°C, to make a zone having a high density of microdefects beneath the denuded zone. This microdefect-rich zone can absorb metallic impurities. The gettering technique has been proved to be practical and effective for some kinds of MOS LSI products.

3. Solid-Phase Epitaxy of Silicon Film on Sapphire

Although most of MOS LSI are produced on Si wafers, thin silicon film on sapphire (SOS) has long been attractive for circuits engineers. This is simply because MOS transistors, diffused layers and metallized layers on SOS consume negligible spurious charging capacitances or negligible ineffective power and because the isolation among transistors occupies smaller area and, in particular, no p- or n-well isolation is necessary for CMOS structures. In the case of thinner films than 0.5 μm, however, SOS has disadvantages of low electron mobility (about half the bulk value) and appreciable current leakage (three-order higher than in the bulk) that result from a large number of defects in the film, in particular, at the silicon and sapphire interface region.

Recently, a new technique to improve remarkably these poor characteristics was devised. It is a kind of solid-phase epitaxy (SPE) or a regrowth of amorphized silicon films.[14] What follows is an example of experiments to show the effectiveness of this technique.

First, to a (001)-oriented epitaxial Si film of 0.3 μm thick on a ($\bar{1}012$) sapphire, which was kept cool by liquid fleon, 190 keV Si ions (penetration depth, $R_p = 2912$ Å) of $1 \times 10^{15}/cm^2$ were ion-implanted in order to amorphize the deep Si layers down to the interface, and then a thermal annealing at 1000°C for 20 min in nitrogen ambient was done. By this annealing, a regrowth of good single-crystalline phase proceeds from the surface region to the interface (single SPE). Next, a shallow Si ion implantation (of $2 \times 10^{15}/cm^2$ at 100 keV, $R_p = 1469$ Å), followed by the annealing, was performed to make a second regrowth from the inner region toward the surface (double SPE). Reprica electron micrographs taken at the surface and at a layer of 0.2 μm in depth showed that the density of stacking-fault based planar defects was drastically decreased by two-order of magnitude. The n-channel transistors made on the 0.3 μm-thick double SPE SOS were certainly proved to show much lower leakage current (see Fig. 4(a)) and much higher electron mobility (see Fig. 4(b)) than those on as-grown SOS.

The electron drift mobility obtained is nearly the theoretical limit of about 80 percents of the bulk value. The limit arizes from that a significant compressive strain in the Si film on sapphire splits the degenerated energy of the k_x, k_y, k_z valleys to up the k_z valleys in the [001] direction and down the k_x and k_y valleys, which leads to an increase in the conductivity effective mass of electrons.[15]

It should be noted that the drift velocity of the electrons in the inverted surface channel tends to saturate when the drain electric field is high enough as in the case of shorter channel length than 2 μm in VLSI. A preliminary experiment[16] showed that under the electric field stronger than 2×10^4 V/cm, the saturating drift velocity of electrons on SOS is nearly same as that on the bulk. This means that one of the disadvantages in SOS will be eliminated in VLSI with 1 μm or less geometry.

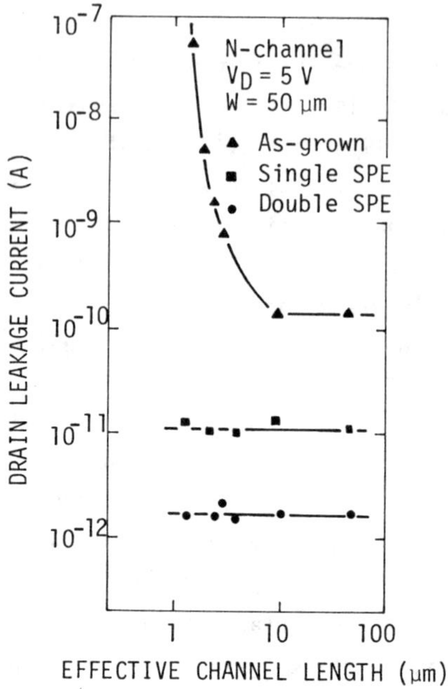

Figure 4(a). Drain leakage current of n-channel transistors on SOS against channel length. After Yoshii (Ref. 14).

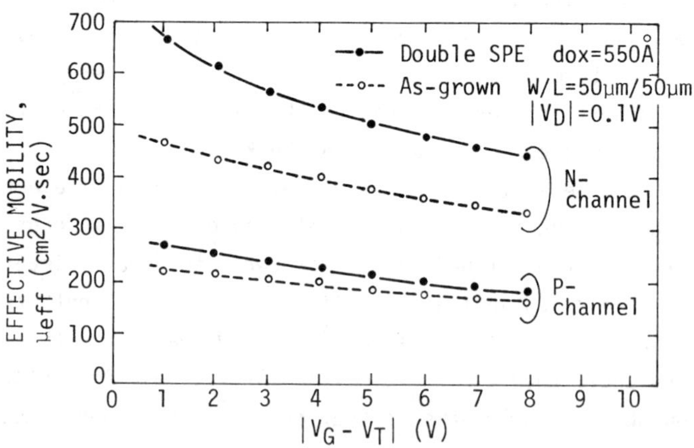

Figure 4(b). Effective mobility of n- and p-channel transistors on SOS as a function of gate voltage, V_G. V_T is the threshold voltage. After Yoshii (Ref. 14).

4. Scaling Down of MOS Transistor Dimensions and VLSI

In 1972 to 1974, Dennard et al. presented in detail how to miniaturize an MOS transistor to one micrometer or less geometry, hoping to increase the integration density and the device performance. They succeeded in showing that properly scaled-down devices give rise to proper circuit performances.[17] Overall scaling results are listed in Table II. The decrease in channel length, channel width, junction depth, gate-oxide thickness, and drain, threshold and gate voltages by K (K is unitless constant) and the increase in substrate impurity concentration by K, should exhibit an I-V characteristics similar to the original one at a reduced level. They demonstrated experimentally the effectiveness of the scaling method.[17] Table III(a) and (a) describe the device parameters of an original transistor and a scaled-down ($K = 5$) transistor, respectively. I_D vs. V_D characteristics for the scaled structure is almost entirely similar to the characteristics at the reduced level by the factor of 5 of the original transistor.

It is noted that the threshold voltage is not exactly scaled by the scalings of the oxide thickness and impurity doping level, but it is practically possible by adjusting the surface impurity concentration by ion implantation. Extensive use of ion implantation provides better overall design of reduced-size transistors. An improved design in this sense is given in Table III(c) which is arranged after Dennard et al.[17] Instead of increasing substrate doping level, (i) a substrate of somewhat lower doping level is used and only surface region is heavily ion-implanted to the extent of the depletion-layer width at the turn-on of the device, and (ii) shallower n^+ junctions comparable to the ion-implanted depth are formed by a heavy ion implantation. The former design (i) is very effective in reducing the substrate bias sensitivity that is a threshold-voltage shift when the source junction is biased above the ground potential. The latter (ii) is effective to avoid punch-through at high drain voltage, because the depletion layers under the source and drain electrodes extend much deeper in the lightly doped substrate. The unfavorable parasitic capacitances are also reduced by this structure.

Table II. Scaling methods of MOS transistor with dimensionless scaling constant K: (a) device parameters, (b) circuit parameters and (c) interconnection line parameters. After Dennard et al.[17)]

Parameter	Scaling Factor
(a) Device parameters	
Channel length L	1/K
Channel width W	1/K
Gate oxide thickness t_{ox}	1/K
Diffused junction depth r_j	1/K
Depletion layer thickness y_d	1/K
Impurity concentration in substrate N	K
Voltage V_D, V_T, V_G	1/K
(b) Circuit parameters	
Current I	1/K
Capacitance $C = A/t$	1/K
Delay time/cirucit VC/I	1/K
Power dissipation/circuit VI	$1/K^2$
Power density VI/A	1
Electric field	1
(c) Interconnection line	
Line resistance $R_L = L/Wt$	K
Voltage drop IR_L/V	K
Response time (delay) $R_L C$	1
Current density I/A	K

Table III. Experiments of scaling down of an n-channel transistor. After Dennard et al.[17)]

Parameter	(a) Original	(b) Scaled (K = 5)	(c) Improved design
t_{ox} (Å)	1000	200	350
L (μm)	5	1	1
W (μm)	5	1	1
x_j (μm)	2.5	0.5	0.2
N_A (atoms/cc)	5×10^{15}	2.6×10^{16}	7.5×10^{15}
N_A in the channel	—	—	—
V_{sub} (V)	-7	-1	-1
V_D (V)	20	4(3)	4
V_T (V)	2.1	0.44	1.1
ϕ_s (V)*	0.65	0.73	

* Band bending in Si at the onset of strong inversion.

These improvements make it possible to increase slightly the gate oxide thickness from 200 to 350 Å to ease general difficulty of the formation of thin oxide, eventhough this sacrificies the gate field.

It should be further noted that the subthreshold current in the weak inversion region of a turning-on transistor cannot be scaled down unless the unrealistic scaling down of the operating temperature is assumed. This fact would be serious for the magnitude of a current switch between "1" and "0" states in reduced-size memory devices.

In general, when the channel length is shortened, say, less than 2 μm, the threshold voltage tends to decrease and to be sensitive on the channel length. This is because the necessary depletion layer charges to turn on the device, that is, to give the surface inversion are not all in the depletion layer under the gate, but the fringe effect of the edges of the source and drain cannot be neglected in the short channel. This is called "short-channel effect". Poon and Yau proposed a practical model effective to analyze the effect.[18)] More exactly, two-dimensional analyses are needed.[19-21)]

In recent years, the theoretical and experimental studies have been extensively performed to obtain the optimum design of small-size transistors in submicrometer range for VLSI in late 1980's. For this demand, several sophisticated process and device simulation techniques have been explored, although some adjusting parameters still remain to fit experimental data. In what follows, a typical result of advanced studies[22)] will be presented.

An n-channel transistor with a channel length of 0.5 μm was fabricated on a highly resistive (50 Ωcm), p-type substrate. The gate oxide is 150 Å. Two boron implants were employed to control the doping profiles: the first deep implant to prevent the punch-through and the second shallow implant to control the threshold voltage. The shallow source and drain junctions of 0.23 μm in depth were provided by arsenic ion implantation through a thin oxide, followed by annealing at 1000°C. The doping profiles were accurately calculated and then the current-voltage characteristics were calculated based on a two-dimensional analysis.

Figure 5(a) shows measured and calculated subthreshold current characteristics for three different channel lengths. It is evident that the ion implantations were well optimized as predicted by a process simulation and the long-channel behavior is well preserved for the 0.5 μm channel. Figure 5(b) is another example of the subthreshold currents observed on a device with a different ion-implantation

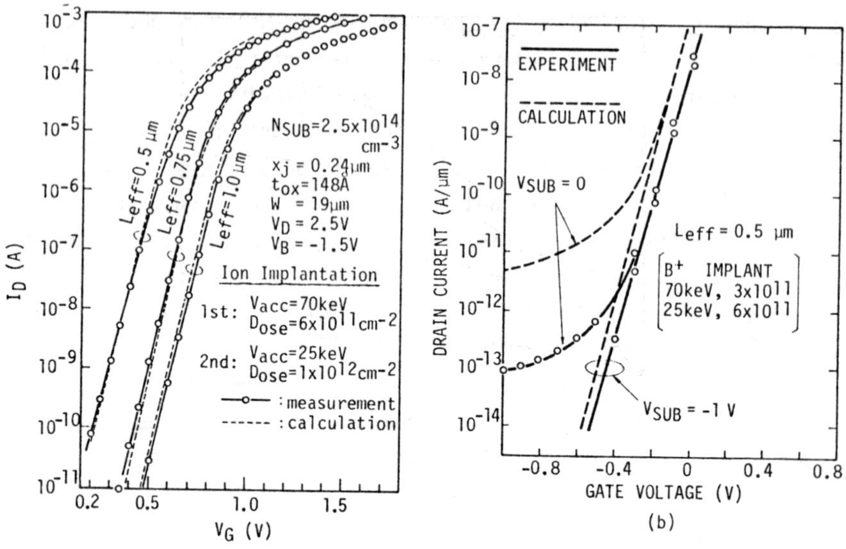

Figure 5. Measured and calculated subthreshold currents as a function of gate voltage: (a) Optimized channel implant conditions. (b) Deep implant dose is somewhat off the optimum. (Ref. 22).

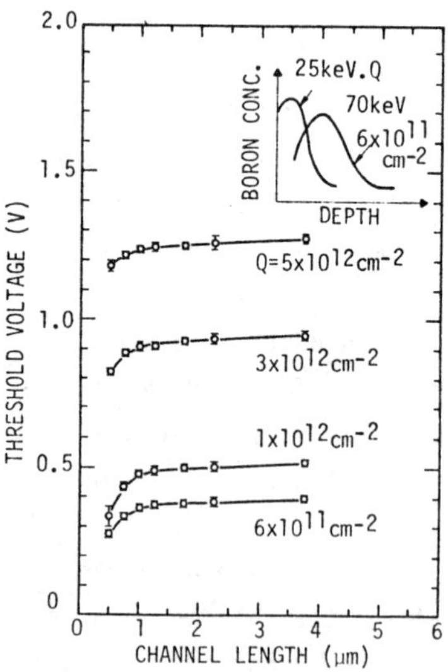

Figure 6. Effect of shallow channel implants on channel length dependence of threshold voltage. ($V_D = 2.5$ V, $V_{sub} = 0$ V. (Ref. 22).

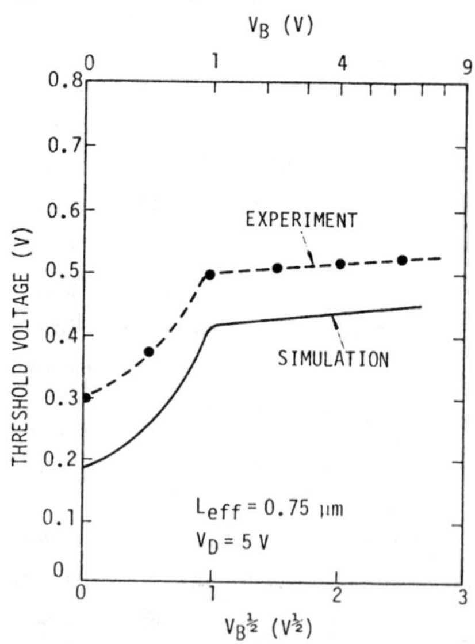

Figure 7. Measured and calculated threshold voltage shifts as a function of substrate bias voltage. (Ref. 22).

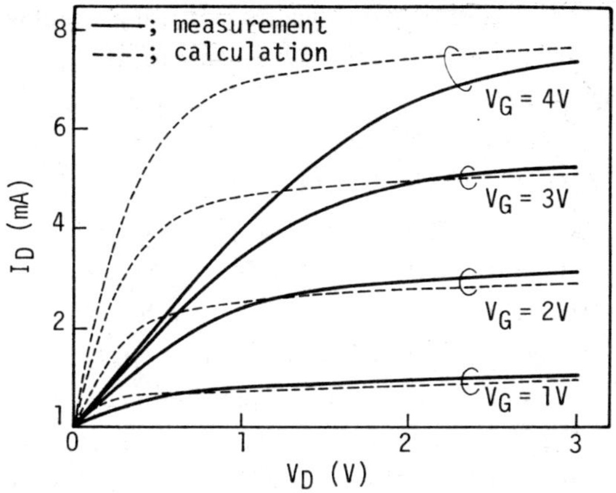

Figure 8. Measured and calculated drain characteristics. (L = 0.5 μm, W = 19 μm). (Ref. 22).

condition, and shows an increase in leakage current with zero substrate bias. When the deep-implant dose was increased to $6 \times 10^{11}/cm^2$, the leakage disappeared. Thus, the leakage is very sensitive to the impurity profile. The short-channel effects are seen in Fig. 6, in which how effective is the channel implant is clear. The sensitivity of threshold voltage to the substrate bias is displayed in Fig. 7. The data were for a device with 0.75 μm channel length. This figure indicates that when a substrate bias of -1 V is applied, the threshold voltage is insensitive to a further increase in the bias. Finally, Fig. 8 shows again I_D vs. V_D curves with the gate voltage as a parameter. The large difference observed between calculation and measurement is due to a large series resistance of the shallow source and drain electrodes (about 50 Ω/□). Practically, this degradation can be successfully eliminated by introducing a new technology which simply plates Pt-silicide on the source, drain and gate electrodes in a self-aligning manner.[22)]

More recently, even an n-channel transistor with a channel length as short as 0.15 μm, a junction depth of 0.1 μm and a gate-oxide thickness of 160 Å was ambitiously fabricated.[23)] Good drain-current and subthreshold-current behaviours similar to those of long-channel devices were successfully demonstrated. This small device was fabricated by an electron-beam lithography for pattern definition.

Although the electron-beam lithography is not very practical yet, the success of this very small working device enables us to be hopeful for the future of super VLSI.

Looking back to Table II(b) and (c), the dimensional scaling leads also to good scaling of other circuit parameters: (i) all circuit elements including interconnection lines will have reduced capacitances by a factor of K, (ii) the delay time of each circuit is reduced by K, (iii) the power dissipation of each circuit is reduced by K^2, (iv) thus, the power-delay product which is a key figure of merit for all circuit units is reduced by K^3, and (v) the power density remains constant. However, the down scaling involves a few problems. For example, the time constant of a transmission line, R_LC is unchanged by scaling, and thus, if a long signal propagation over long lines is involved, higher switching speed is not realized.[24] Also, the increase in current density by K would cause a reliability problem. To avoid these disadvantages, multi-level interconnection technology is desired which makes it possible to widen power bus lines and to afford the use of over-layer metal lines instead of heavily doped signal lines in the substrate.

In order to integrate a tremendous number of these scaled-down devices, innovative production technologies have to be explored for every step of a number of processings. Of particular interests are (i) fine-line lithography with high accuracy and precise registrations among many succeeding photomask layers to control, for example, the threshold voltage to be less than ±0.1 V, (ii) growth of pinhole-free thin oxide, and (iii) anisotropic dry etching to define exactly the edges of all electrodes and overlayers of insulators and metals. Circuit designers always confront the tolerance limited by available technologies, which inevitably determines circuit performances. We have discussed only the channel length, but the channel width is also important for "narrow-channel effect",[25] which is closely related to isolation structures among adjacent transistors. When going down to 1 μm range, conventional LOCOS (local oxidation of silicon) process, utilizing Si_3N_4 layers on the surface, occupies an isolation area larger than the transistor size. Answering to this problem, several new isolation technologies are widely under development.

Finally it is noted that most of the scaling-down studies have been done on the n-channel devices, and in contrast, the study of scaled-down p-channel transistors is behind. The p-channel transistors, which were important in late 1960's (see Table I), are increasingly important for CMOS LSI that need isolation of p- and n-type circuits area except SOS, but are very attractive for low-power consumption. Even in the transistor sizes in most of current advanced VLSI products, the drain field is high enough to put the drift velocity of channel electrons to saturate by optical phonon scattering. The saturation value of hole drift velocity is nearly same as that of the electrons.[26] Also, hot holes induced by impact ionization in the drain accelerating field have less chance to surmount the potential barrier at the oxide-semiconductor interface into the oxide than hot electrons, which is an interesting problem to be discussed in next section. One of the difficulties in making small p-channel transistors is to make shallow junctions unless temperature is much decreased, because of large diffusion coefficient of boron. More effort to study the scaled-down p-channel transistors[27] is hoped.

5. Hot-Carrier Effects and VLSI

When the drain field is high, non-thermal hot carrier would be generated by impact ionization. This really occurs and gives rise to carrier injection into the gate oxide and into the substrate. Whereas the carrier injection into the gate can be intentionally applied to a special memory device, programmable read-only memory, the hot carrier effects are extremely important concerning degradation of device performance or reliability in VLSI. In this section, some of these and other competing problems will be outlined briefly.

5.1 Avalanche Electron Injection into the Oxide and EPROM

About a decade ago, it was found that, when the drain electrode of a p-channel transistor is applied carefully by just a breakdown voltage, the avalanche voltage gradually increases.[27] This phenomenon is called "walk out". This is due to charging of the traps, otherwise neutral, in the oxide above the edge of the drain by trapping negative charges which were injected hot electrons from the drain avalanche. Then, the line of electric force from the gate to the semiconductor surface is weakened, and now, larger drain voltage is necessary to breakdown the drain junction. If the gate oxide is carefully made to have a negligible density of trapping centers and if the gate is floated from external electrodes, the injected electrons are trapped in the floating gate, which in turn induce counter charges on the channel region or shift the threshold voltage. In other words, this turns a normally-off state of a p-channel transistor to a normally-on state. This is the principle of floating-gate, avalanche-injection type, electrically programmable read-only memory, EPROM.[29,30] The erasure of written memory is readily done by photoemission of the electrons from the floating gate by uv irradiation.

In general, the probability of the injection is much larger for the hot electrons than the holes, because the potential barrier height at Si-SiO_2 interface which the hot carriers should surmount is lower for electrons (3.15 eV) than holes (3.8 eV), and also because the optical-scattering mean free path, λ, in Si is longer for electrons (60 Å) than holes (45 Å). Thus, so far as the avalanche injection is a design principle, the p-channel type EPROM is much easier to fabricate than the n-channel type, as was the first 2 k-bit EPROM products. In the n-channel type, hot-hole injection is certainly observed, but the hot electrons can also be injected even against a repelling potential configuration, if it is not too large.

In advanced EPROM's of bit capacities of more than 16 k bits, however, n-channel transistors have been used for the memory cell, because smaller sizes and faster access time are possible. In these n-channel EPROM's, the hot electrons to surmount the barrier are designed to be generated by impact ionization in the channel region near, but not at the drain junction.[31] Theoretical and experimental analyses of the channel hot-electron injection have been reported.[31-39,41] The analyses should have been performed on the electric field configuration, drift velocity saturation, impact ionization efficiency, hot-electron energy and momentum distribution function and injection efficiency.

5.2 Hot-Electron Injection in Scaled-Down Devices

One of the advanced two-dimensional calculations of hot-electron current into the gate and of hot-hole current into the substrate in small size n-channel transistors will be exemplified.[38,39] Basing on a process/device simulator[40] for a configuration of device parameters, the electric field distribution is first calculated, and then the impact ionization process is calculated[38]; The carrier multiplication factor M is:

$$M = 1/(1 - I) \; ; \tag{1}$$

$$I = \int \alpha_n \exp\left(-\int_0^x (\alpha_n - \alpha_p)dx'\right)dx \; , \tag{2}$$

where α_n and α_p are impact ionization coefficients of electrons and holes, respectively, defined by

$$\alpha = A \exp(-B/E) \; . \tag{3}$$

Here E is the electric field strength, and A and B are constants. The generated hole current in a small region of interest is

$$\Delta I_{sub} = (M - 1)J_x \; , \tag{4}$$

and the total substrate current is the integration of ΔI_{sub} throughout the substrate.

Contribution to the injected electron current into the gate oxide is obtained by calculating the fraction of electrons reaching the interface, $\exp(-d/\lambda)$, and the energy- and momentum-dependent probability of escape over the interface potential barrier. Notation d is the distance from the generated point to the interface. The injected current reaching the gate electrode is then calculated, accounting for an appreciable influence of the lateral local electric field as illustrated in Fig. 9(b).[39]

The calculated and measured results of the substrate and gate injection currents are plotted as functions of the gate voltage in Fig. 10(a) and (b), respectively.[39] Note that the gate injection current as small as 10^{-21} A can be measured by the measurable threshold voltage shifts in an EPROM structure.

The results of this kind of study are helpful for better designs and reliability of future high-density EPROM's. On the other hand, the hot-electron injection is one of the most serious constraints in

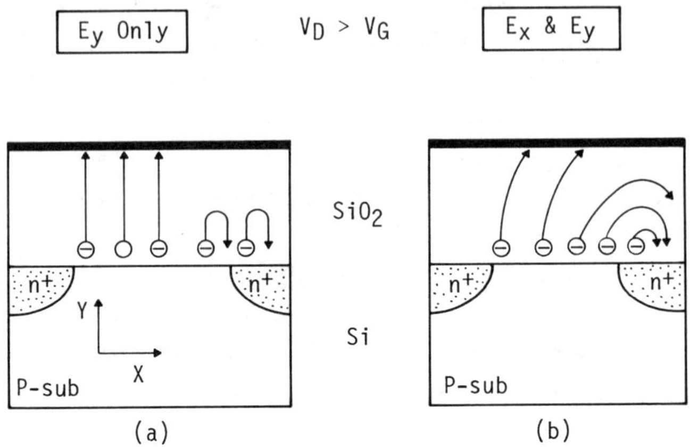

Figure 9. Models of the hot-electron injection into the gate oxide. (a) Only the vertical component of the oxide field is assumed. (b) Two-dimensional calculation model. (Ref.39).

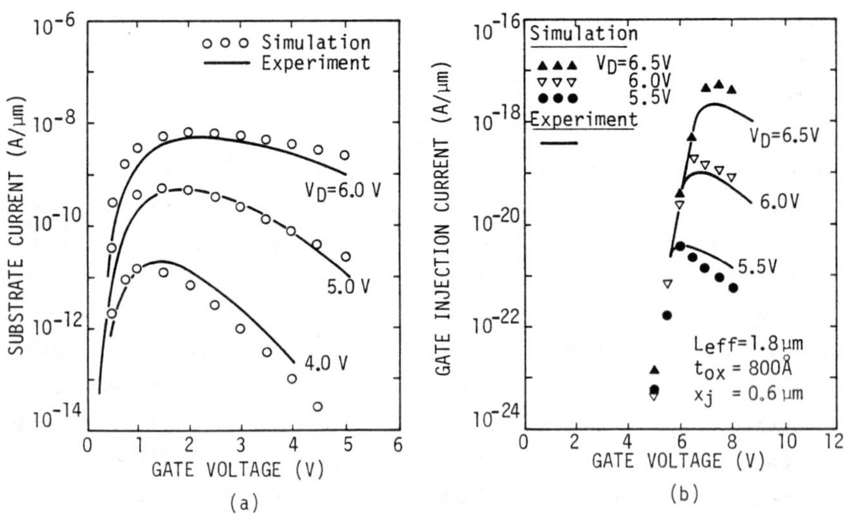

Figure 10. Measured and calculated hot-carrier currents. (a) Injected hole current to the substrate. (b) Injected electron current to the gate. (Ref. 39).

scaled-down devices in VLSI.[42] Surface states of a finite density in the gate oxide may be charged up by trapping the injected electrons. Shift of only one tenth of a volt in the threshold voltage degradates the performance of sophisticated VLSI. Magnitude of the degradation is determined by the drain field, the gate field, the nature and density of the traps and temperature. Extensive experimental data on this subject are available.[43]

One simple solution to cut out this constraint is to replace power supply standard, 5 V, by a reduced supply, say, 3 V or by a reduced internal supply on a chip. Otherwise, new device structures to minimize the hot-carrier generation, the injection and other effects are inevitable. There have been many attempts for this purpose. One is a double-implanted lightly doped drain/source structure, which provides self-aligned n^- regions between the channel and n^+ diffused drain and source.[44] This LDD structure evidently reduces the maximum field intensity near the drain by spreading the high field at the drain. Takeda et al.[45] examined the effects of various new device structures on the hot-electron injection and highest applicable drain voltage: (i) As-P ($n^+ - n^-$) double-diffused drain/source, (ii) P-doped drain, (iii) offset gate separating the gate electrode a bit from the localized peak of the drain electric field and (iv) burried p^+-channel provided beneath the channel inversion layer. The most effective structure was found to be surprisingly the P-doped structure.

5.3 Secondary Generated Minority Carrier

In recent years, two failure modes were found in d-RAM's. One is so-called "soft error".[46] When α particle is passing through the memory devices, generated electrons diffuse to the memory cell and fill an empty (depleted) cell: "1" state to "0". The α particle originates from α decay of uranium or thorium in package material. The α particle energy centered around 5 MeV, extending to 9 MeV, may generate 1.4×10^6 electron-hole pairs in passing 30 μm long. The soft error would be serious for not only d-RAM's, but also all kinds of high-density memory and CCD (charge-coupled device). A variety of designs of the memory structures to minimize the pass has been developed, and simply, a polyimide film coating are added on finished chips before packaging for current d-RAM's.

Another failure mode is the effect of minority carrier generation by secondary impact ionization occurring in the drain depletion region,[36,47] as illustrated schematically in Fig. 11.[47] It was first found in a 16 k-bit d-RAM and an experimental CCD that, when a peripheral large MOS transistor is operated in the saturation region by a high voltage, the memory cells near the transistor start to fail ("1" to "0") and this failure mode extends radially to other adjacent cell arrays.[48] The phenomenon was interpreted, by including further independent experiments, to be due to the diffusion of spurious energetic electrons generated at the depletion region edge (see Fig. 11) by secondary impact ionization caused by accelerated holes that had been generated by the primary impact ionization in the channel near the drain. The generated energetic minority carrier (electrons) may run in the neutral bulk against a repelling field to an empty memory cells to turn on ("0"). The same kind of phenomenon was found to occur in the p-channel devices as well.[49]

Overall results of systematic measurements of primary and secondary impact ionization currents in the n- and p-channel transistors devised in p- and n- wells, respectively, are reproduced in Fig.12.[49] The observed substrate current, I, is given by

$$I = I_0 \exp(-x_w/\lambda_{eff}) \quad , \tag{5}$$

where I is total current injected from the drain depletion edge, x_w well depth and λ_{eff} effective diffusion length of minority carrier. The primary and secondary impact ionization rates, α_1 and α_2, may be defined as

$$\alpha_1 = I_1/I_{DS} \simeq \int \alpha_n \exp[-\int^x (\alpha_n - \alpha_p)dx']dx \quad , \tag{6}$$

$$\alpha_2 = I_2/I_1 \quad . \tag{7}$$

Here, I_{DS}, I_1 and I_2 are, respectively, source to drain current, primary and secondary impact ionization currents, and α_n and α_p denote, respectively, electron and hole impact ionization rates. The values of α_1, α_2 and λ_{eff} are listed in Table IV. It is important to note that majority carrier (electrons) current in the p-channel transistor

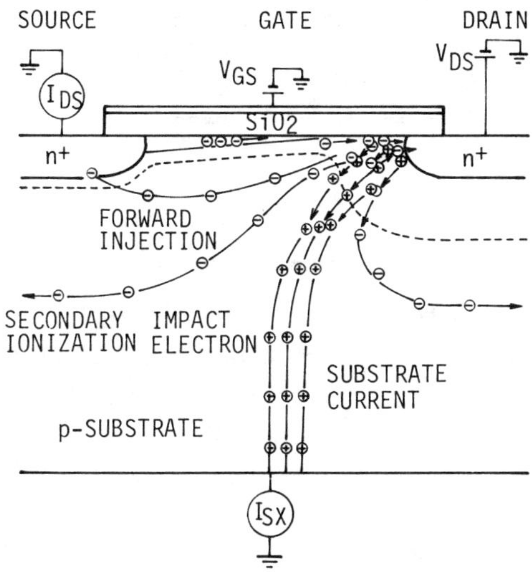

Figure 11. Schematic illustration of channel electrons, primary impact ionization holes and secondary impact ionization electrons in an n-channel transistor. After Matsunaga (Ref. 47).

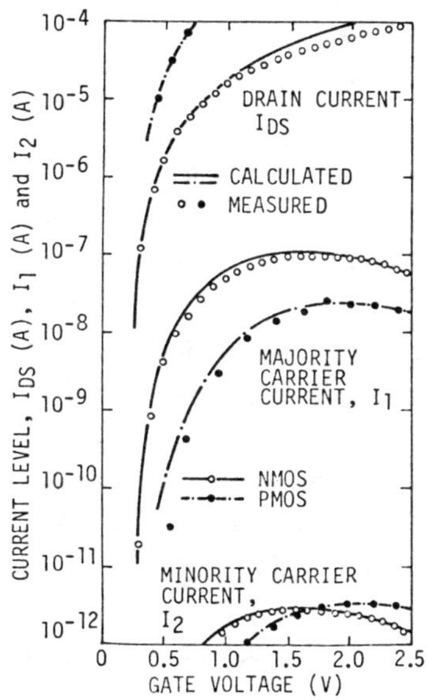

Figure 12. Drain current, primary impact ionization current, and secondary impact ionization current as a function of gate voltage in n-channel and p-channel transistors. After Matsunaga (Ref. 49).

Table IV. Summary of primary (α_1) and secondary (α_2) impact ionization rates and effective diffusion length of minority carrier (λ_{eff}). (Ref. 49).

	α_1	α_2	λ_{eff}
NMOS	5.5×10^{-4}	2.8×10^{-5}	46 μm
PMOS	1.0×10^{-7}	1.0×10^{-4}	45 μm

is one order of magnitude less than the one (holes) in the n-channel transistor. This means that the hot-carrier effect described in the preceding subsection is eased in scaled-down p-channel transistors than n-channel.

Very recently, two new interpretations of the origin of the holding-time degradation in NMOS d-RAM's were reported. One report[50)] claims that the diffused electrons in the substrate originates at a forward-biased source-substrate junction formed due to a resistive potential drop in the substrate caused by the hole current. Another study[51)] found two terms in the minority carrier current: a linear dependence on the hole current in lower region and an exponential dependence in higher region. The mechanism of the latter fact is interpreted to be same as the above mechanism, but the linear dependence is claimed to be due to photon generation in the drain impact ionization zone. Approximately, 2×10^{-5} photons having near-bandgap energy are generated for every hole produced by the impact ionization. Then, these photons travel long distance enough to the adjacent transistors.[51)] Validity of these new interpretations has not been clear enough yet. If the photon generation mechanism truly works, it imposes a new restriction in VLSI.

5.4 Punchthrough and Breakdown

Remaining important constraints in the scaling down of the transistors are source to drain current punchthrough and transistor breakdown at high drain voltages. Punchthrough is a kind of pre-breakdown and occurs when the depletion region extends from the drain to the source and the source potential barrier is lowered for electrons to inject into the depletion region and to the drain. The avalanche breakdown of short-channel transistors results from an ohmic drop caused by the substrate current and a positive feedback effect of a

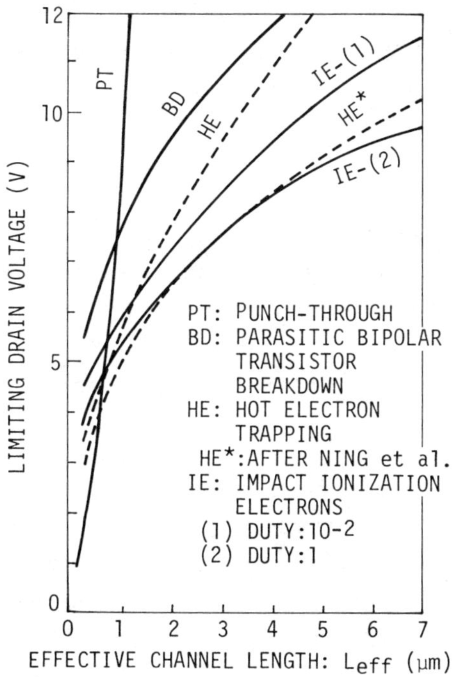

Figure 13. Estimated limiting drain voltages for various important effects as functions of effective channel length. After Matsunaga (Ref. 47).

parasitic lateral bipolar transistor, which involves substrate hole current. Because of the importance of these phenomena, various theoretical analyses have been performed.[44,52-56] Remarkably effective device structures having the channel length of 0.5 μm to ease these constraints have been a p^+-burried channel structure[57] and a sophisticated p-pocket formation under the lightly doped drain/source (LDD) configuration[58].

5.5 Maximum Working Drain Voltage Limited by Multiple Effects

In Fig. 13 are shown estimated maximum drain voltages limited by various constrains discussed above in conventional n-channel transistors.[46] Looking at the effective channel length in the region of 1 μm and submicrometers, all limiting mechanisms are competing with one another. Thus, further systematic study of these effects including the case of p-channel transistors[59] and development of innovative device structures like those cited above to overcome the constraints are essentially important for the basis of future VLSI with submicrometer feature sizes.

6. Concluding Remarks

We have reviewed recent studies of basic problems in Si material technology and in miniaturizing MOS devices. We have many other basic subjects concerning processing technologies such as oxidation and nature of thin oxide, pattern lithography and physical etching, and metallized contacts and interconnection. Unless total technology development is pursued, optimum high-density VLSI featuring the device size in every period would be never realized. Nevertheless, the study of basic subjects such as those we have picked up will go further on much smaller devices of channel lengths shorter than 0.5 μm to expose new problems to be solved in advance before economical design and production technology of integration will be developed, and even to find a fundamental limit in the size of the conventional device structures.

Acknowledgments

The author acknowledges a number of his colleagues, in particular, Shin. Takasu, S. Kohyama, T. Yoshii, J. Matsunaga, T. Shibata , M. Wada and H. Tango for discussions of their studies.

References

1) M. Watanabe, T. Usami, H. Muraoka, S. Matsuno, Y. Imanishi and H. Nagashima, Semiconductor Silicon 1981, ed., H. R. Huff, R. J. Kriegler and Y. Takeishi, The Electrochemical Society, 1981, p. 126.
2) T. Suzuki, N. Iwasa, Y. Okubo and K. Hoshi, ibid., p. 90.
3) K. Hoshikawa, Jpn. J. Appl. Phys. 21, L545 (1982).
4) S. Takasu, H. Otsuka, N. Yoshihiro and T. Oku, Proc. 12th Conf. Solid State Devices, Tokyo, 1980, Jpn. J. Appl. Phys., 20 (1981), Suppl. 20-1, p. 25.
5) H. Muraoka, S. Takasu, M. Watanabe, Y. Matsushita and M. Ogino, Toshiba Review, 36, 603 (1981), (in Japanese).
6) T. Y. Tan, E. E. Gardner and W. K. Tice, Appl. Phys. Lett., 30, 175 (1977).
7) Y. Matsushita, S. Kishino and M. Kanamori, Jpn. J. Appl. Phys.,

19, L101 (1980).
8) S. Kishino, Proc. 15th Int. Conf. Physics of Semiconductors, Kyoto, 1980, J. Phys. Soc. Jpn., 49 (1980), Suppl. A. p. 49.
9) S. Kishino, Proc. 13th Conf. Solid State Devices, Tokyo, 1981, Jpn. J. Appl. Phys., 49 (1982), Suppl. 21-1, p. 105.
10) Y. Matsushita, J. Crys. Growth, 56, 516 (1982).
11) K. Yamamoto, S. Kishino, Y. Matsushita and T. Iizuka, Appl. Phys. Lett. 36, 195 (1980).
12) K. Nagasawa, Y. Matsushita and S. Kishino, Appl. Phys. Lett., 37, 622 (1980).
13) S. Kishino, K. Nagasawa and T. Iizuka, Jpn. J. Appl. Phys., 19, L466 (1980).
14) T. Yoshii, S. Taguchi and T. Inoue, Proc. 13th Conf. Solid State Devices, Tokyo, 1981, Jpn. J. Appl. Phys. 21 (1982), Suppl. 20-1, p. 175.
15) Y. Ohmura, Proc. 7th Conf. Solid State Devices, Tokyo, 1975, Jpn. J. Appl. Phys. 15, 1976, Suppl. 15-1, p. 233.
16) H. Tango, private communication.
17) R. H. Dennard, F. H. Gaensslen, H. N. Yu, V. L. Rideout, E. Bassous and A. R. Le Blanc, IEEE J. Solid-State Circuits, SC-9, 256 (1974).
18) L. D. Yau, Solid-State Electr. 17, 1059 (1974).
19) M. S. Mock, ibid, 16, 601 (1973).
20) D. P. Kennedy and P. C. Murley, IBM J. Res. Develop. 17, 1 (1973).
21) T. Toyabe, K. Yamaguchi, S. Asai and M. S. Mock, IEEE Trans. Electr. Dev., ED-25, 825 (1978).
22) T. Shibata, K. Hieda, M. Sato, M. Konaka, R. L. M. Dang and H. Iizuka, ibid, ED-29, 531 (1982).
23) W. Fichtner, R. K. Watts, D. B. Fraser, R. L. Johnston and S. M. Sze, IEEE Elecr. Dev. Lett. EDL-3, 412 (1982).
24) K. C. Saraswat, F. Hohammadi, IEEE Trans. Electr. Dev., ED-29, 645 (1982).
25) R. Dang, N. Shigyo, T. Wada, S. Onga and M. Konaka, Proc. Int. Conf. Circuits and Computers (ICCC 82), Oct., 1982, p. 286.
26) T. Sato, Y. Takeishi, H. Tango, H. Ohmura and Y. Okamoto, J. Phys. Soc. Jpn., 31, 1846 (1971). Other recent studies are referred to K. K. Thornber, J. Appl. Phys. 51, 2127 (1980).
27) Y. El-Mansy, IEEE Trans. Electr. Dev. ED-29, 567 (1982).
28) H. Hara, Y. Okamoto and H. Ohnuma, Jpn. J. Appl. Phys., 9, 1103 (1971).
29) D. Frohman-Bentchkowsky, IEEE J. Solid State Circuits, SC-6, 301 (1971).
30) H. Iizuka, T. Sato, F. Masuoka, K. Ohuchi, H. Hara, H. Tango, M. Ishikawa and Y. Takeishi, Proc. 4th Conf. Solid State Devices, Tokyo, 1972, Oyo Butsuri (Jpn. J. Appl. Phys.) 42 (1973), Suppl., p. 158.
31) T. Hagiwara, E. Takeda, M. Horiuchi, R. Kondo and Y. Ito, Proc. 8th Conf. Solid State Devices, Tokyo, 1976, Jpn. J. Appl. Phys. 16 (1977), Suppl. 16-1, p. 211.
32) P. E. Cottrel, R. R. Troutman and T. H. Ning, IEEE Trans. Electr. Dev. ED-26, 520 (1979).
33) T. H. Ning, P. W. Cook, R. H. Dennard, C. M. Osburn, S. E. Schuster and H. N. Yu, ibid, ED-26, 346 (1979).
34) S. Tanaka and M. Ishikawa, ibid, ED-28, 1190 (1981).
35) B. Eitan and D. Frohman-Bentchkowsky, ibid, ED-28, 328 (1981).
36) S. Kohyama, T. Furuyama, S. Mimura and H. Iizuka, Proc. 11th Conf.

Solid State Devices, Tokyo, 1979, Jpn. J. Appl. Phys., 19 (1980), Suppl. 19-1, p. 85.
37) M. Wada, S. Mimura and H. Iizuka, IEEE Int. Electron Devices Meeting, Techn. Digest, p. 38 (1980).
38) M. Wada, T. Shibata, M. Konaka, H. Iizuka and R. L. M. Dang, IEEE Int. Electron Devices Meeting, Techn. Digest, p. 223 (1981).
39) M. Wada, T. Shibata, H. Iizuka and R. L. M. Dong, 1982 Symp. VLSI Techn., Oiso, 1982, Digest Techn. Papers, p. 102.
40) S. Onga, M. Konaka, T. Taniguchi, H. Iwai and R. L. M. Dang, Semiconductor Silicon 1981, ed., H. R. Huff, R. J. Kriegler and Y. Takeishi, The Electrochemical Society, 1981, p. 1020.
41) S. Tam, P. K. Ko, C. Hu and R. S. Muller, IEEE Trans. Electr. Dev., ED-29, 1740 (1982).
42) D. K. Ferry, J. R. Barker and H. L. Grubin, ibid, ED-28, 905 (1981).
43) H. Matsumoto, K. Sawada, S. Asai, H. Hirayama and K. Nagasawa, ibid, ED-28, 923 (1981).
44) S. Ogura, P. J. Tsang, W. W. Walker, D. L. Critchlow and T. Shepard, ibid, ED-27, 1359 (1980).
45) E. Takeda, H. Kume, T. Toyabe and S. Asai, 1981 Symp. VLSI Techn., Hawaii, 1981, Digest Techn. Papers, p. 22. Also, IEEE Trans. Electr. Dev., ED-29, 611 (1982).
46) T. C. May and M. H. Wood, Proc. 1978 Int. Reliability Phys. Symp., San Diego, 1978, p. 33.
47) J. Matsunaga, S. Kohyama, M. Konaka and H. Iizuka, Proc. 11th Conf. (1979 Int.) Solid State Devices, Tokyo, Jpn. J. Appl. Phys., 19 (1980), Suppl. 19-1, p. 93.
48) T. Furuyama, K. Ohuchi and S. Kohyama, IEEE Trans. Electr. Dev., ED-26, 1684 (1979).
49) J. Matsunaga, H. Momose, H. Iizuka and S. Kohyama, IEEE Int. Electron Devices Meeting, Techn. Digest, p. 736 (1980).
50) B. Eitan, D. Frohman-Bentchkowski and J. Shappir, IEEE Trans. Electr. Dev., ED-28, 1515 (1981).
51) S. Tam, F. C. Hsu, P. K. Ko, C. Hu and R. S. Muller, IEEE Electr. Dev. Lett., EDL-3, 376 (1982).
52) T. Toyabe and S. Asai, IEEE Trans. Electr. Dev., ED-26, 453 (1979).
53) R. R. Troutman, ibid, p. 461.
54) F. C. Hsu, P. K. Ko, S. Tam, C. Hu and R. S. Muller, ibid, ED-29, 1735 (1982).
55) E. Sun, J. Moll and J. Berger, IEEE Int. Electron Devices Meeting, Techn. Digest, p. 478 (1978).
56) B. Eitan and D. Frohman-Bentchkowski, IEEE Trans. Electr. Dev., ED-29, 254 (1982).
57) H. Sunami, K. Shimoyanagi and N. Hashimoto, ibid, p. 607.
58) S. Ogura, C. Codella, N. Rovedo, J. Shepard and J. Riseman, 1982 Symp. VLSI Techn., Oiso, Digest Techn. Papers, p. 42.
59) Y. A. El-Mansy, 1981 Symp. VLSI Techn., Hawaii, Digest Techn. Papers, p. 16.

THIRTY YEARS OF SEMICONDUCTOR PHYSICS
— A BRIEF HISTORICAL SKETCH AFTER "TRANSISTOR" —

Yasutada Uemura

Department of Applied Physics
The Science University of Tokyo
1-3 Kagurazaka, Shinjuku, Tokyo, Japan

A brief historical sketch of an overview on the developments of modern semiconductor physics after the discovery of transistor action is described by placing particular emphasis on those in Japan. The thirty years are divided into three decades. The first decade (1950-1959) is represented by the heading of "The dawn of modern semiconductor physics", the second one (1960-1969) by "Extension and sophistication" and the third one (1970-1980) by "Sign of transformations".

1. Introduction

It was a great pleasure and also a great honor for me that there should be a symposium in my honor at the occasion of my retirement from the University of Tokyo this spring. The symposium covered several topics on the recent semiconductor physics, the field to which I have paid my continual interests in these thirty years. The editors kindly invited me to write the last chapter for this book which is mainly based on the lectures given at the symposium. Therefore I will describe a brief historical sketch on the developments of semiconductor physics in these thirty years by placing particular emphasis on those in Japan. I hope that the sketch will help us to understand the general background of recent topical activities discussed in this book.

Everyone can agree that the discovery of transistor action and the invention of transistor in 1947 were one of the most epoch-making events in the history of semiconductor physics. The discovery initiated to open not only a new promising field of electrical engineering, now called "solid state electronics", but also a very important development in the field of basic research now called "solid state physics". The historical sketch in this chapter covers the period after the discovery of transistor action in 1947 to the present, and the phrase "these thirty years" means the period of roughly from 1950 to 1980.

At present we have "The International Conference on the Physics of Semiconductors" (I.C.P.S.) sponsored by I.U.P.A.P. every two years. Table 1 is a chronological table of every I.C.P.S. indicated by the name of its host city. The first Reading Conference was held in July 1950 in England and it was organized by Profs. R. W. Ditchburn and N. F. Mott. The second one was held in 1954 at Amsterdam, and after that I.C.P.S. were held every two years. We are going to have the 16th I.C.P.S. at Monpellier in September 1982. As to the I.C.P.S. in Japan, the 8th and 15th conferences were held at Kyoto, and gave verious stimuli to the research activities of semiconductor physics in Japan. In the following sections, I will describe the historical sketches divided into three decades. The first decade (1950-1959), the second decade (1960-1969) and the third decade (1970-at present) are characterized

TABLE 1.

Host city	Year	
	1950	
READING		
AMSTERDAM		THE DAWN OF MODERN SEMICONDUCTOR PHYSICS
GARMISCH		
ROCHESTER		
PRAGUE	1960	
EXETER		
PARIS		EXTENSION AND SOPHISTICATION
KYOTO		
MOSCOW		
CAMBRIDGE	1970	
WARSAW		
STUTTGART		SIGN OF TRANSFORMATIONS
ROME		
EDINBURGH		
KYOTO	1980	

by their own headings on the right of Table 1. The reasoning for the choice of these headings for each decade will be given in the following sections which describe typical characteristics of the research activities in these decades. The papers published in every "Proceedings of I.C.P.S." were used as the basic references to grasp the international trends of semiconductor physics. I did not make any systematic reference to papers published in other original journals for this article.

As I have already mentioned, an emphasis is placed on the history of research activities in Japan. The recent remarkable growth of semiconductor industries in Japan is drawing a world-wide attention today. Therefore, it may be interesting to show how Japan's activities in basic researches have also developed with the growth of industrial activities. For this purpose, two graphs, Fig. 1 and Fig. 2 are presented prior to a detailed description of historical sketches. I am very much indebted to Dr. Akiko Natori for her generous assistance and effort to arrange data for these graphs.

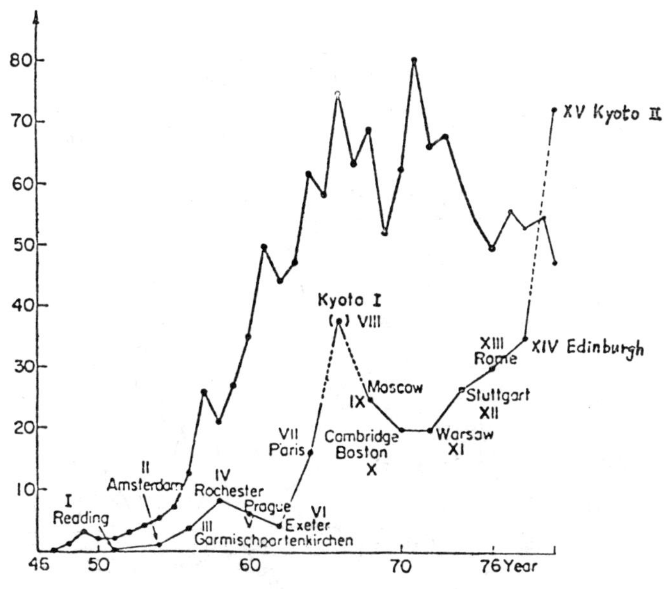

Fig. 1 Number of papers on semiconductor physics published in J.P.S.J. (Journal of Physical Society of Japan) in every year. Number of papers published in the "Proceedings" of every I.C.P.S. by a Japanese physicist as the first author. The conferences are indicated by the names of host city.

The upper curve in Fig. 1 shows the number of original papers on semiconductor physics published in J. Phys. Soc. Japan in each year since 1947. The lower curve in Fig. 1 shows the number of papers published in every "Proceedings of I.C.P.S." by Japanese physicists as the first authors. As you can see, both curves are plotted against the scale of year so that the curves suggest some indication of the growth rate on basic semiconductor physics research in Japan.

In Fig. 2, the number of papers plotted in the upper curve of Fig. 1 is classified by materials with which each paper was mainly concerned. Figs. 2(a) and 2(b) are the numbers of papers on the bulk properties of single crystals of Ge and Si respectively. Fig. 2(c) is for papers concerned with the properties of junctions, surfaces and

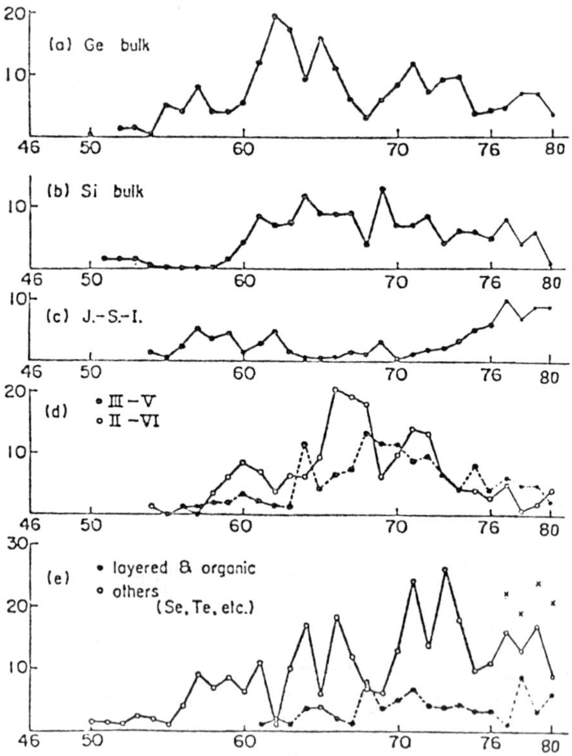

Fig. 2 The papers on semiconductor physics published in J.P.S.J. are classified by materials with which each paper was mainly concerned. The number of papers in each category was plotted against the year.

interfaces. For example, papers on p-n junctions of Ge were the main contributors to the curve in the earlier stage of these thirty years, while papers on two dimensional electrons in MOSFET took a major part in the recent upward trend of the curve as shown in Fig. 2(c).

The papers on III-V compounds such as InSb and GaAs and on II-VI compounds such as CdS and ZnO were plotted in Fig. 2(d) by ● and ○ respectively. In Fig. 2(e), the papers on layered and organic semiconductors including graphite were plotted by ●, while papers on the remaining materials including Se and Te etc. were shown by ○.

By glancing at these curves in Fig. 1 and Fig. 2, the reader is supposed to be able to make his own speculative image on the development of semiconductor physics in Japan. I hope that some comments in comparison with the international trends of the field given in the following sections will be useful for every reader in drawing his own conclusion from the curves shown in Fig. 1 and Fig. 2.

2. The Dawn of Modern Semiconductor Physics

The first decade starting in 1950 can be characterized as "The Dawn of Modern Semiconductor Physics" opened by the discovery of Transistor action.

A large single crystal of germanium having very high purity and perfection was the key material of this decade. It was quite reasonable that a motivation to obtain such a high quality sample of germanium came first from a technological need to produce good transistors. However, it was also true that this motivation was very beneficial for the study of basic properties of semiconductors experimentally, because experimental verifications of various theoretical concepts for an idealized perfect crystal were expected to be done only by using samples with quantitatively controlled impurities and defects. For semiconductor physics before the advent of the transistor, the possibility of it being an "exact science" was highly dependent on the technological progress in controlling the so called "structure sensitive" properties of materials.

Czochraski's method of growing crystals was very much improved so that sample temperatures and the pulling rate of growing crystals could be precisely controlled. The invention of the "zone melting method" by R. N. Hall and W. G. Pfann brought us samples having extremely high degree of purification. These developments in material technology in the first half of the decade made single crystals of germanium to be the key material of modern solid state physics, and later these development were extended to cover various other kinds of samples not only of semiconductors but also of insulators and metals.

Every semiconductor physicist knows that observations of the band gap (thermal and optical) and of the temperature dependence of conductivity and Hall coefficient are the starting points of semiconductor research. The results were interpreted by the so-called "Wilson model" of semiconductors proposed in 1931.

As to the optical studies on the band gap of Ge, one can find a report by H. Y. Fan and M. Becker at the 1950 Reading Conference. After the earlier results of band calculation for Ge which had been done mainly by F. Herman by 1954, their experimental identification of both direct and indirect gaps referred to the symmetrical points in the Brillouin zone was almost the same as those we know at present.

The detailed experimental studies on the temperature dependence of conductivity and Hall coefficient in n-type germanium were reported by P. P. Debye and E. M. Conwell in 1954. It was presumably the first reliable quantitative experimental results on the thermal band gap, binding energies of carriers at donors and even on the value of the density state mass of conduction electrons of semiconductors. The observation of anisotropic effects in magnetoresistance by G. L. Pearson and H. Suhl in 1951 had already suggested the many valley structure of the conduction band of germanium, and this structure was also verified by the piezoresistance effect by G. L. Smith in 1954.

As to the valence band of germanium, the light and heavy hole bands were recognized as well as the so-called split-off band by the optical absorption induced by the injected holes in these bands. A paper on these findings was presented by H. B. Briggs and R. C. Fletcher in 1953. In 1954, the transport properties of holes in germanium was observed by R. K. Willardson, T. C. Harman and A. C. Beer in which an anomalous magnetic field dependence of Hall effect was attributed to the presence of two kinds of holes having light and heavy effective masses.

Thus one can find that almost all of the basic electronic properties of bulk single crystal of germanium had been studied by 1954.

Reflecting these developments in various experimental studies, many interesting papers were presented to the 2nd I.C.P.S. at Amsterdam in 1954 as if to signal the flowering of germanium studies. In this conference, beautiful cyclotron resonance experiments to determine the effective mass tensor of carriers were reported by two groups. A. F. Kip et al. in Berkeley and B. Lax et al. in M.I.T. I suppose that there are many solid state physicists who still keep their fresh and exciting impressions of the announcement the success of the experiments at first hand. It was really a typical example of modern solid state experiments, because it was based on the very modern technologies of that time, namely those for high quality samples, for liquid helium and for microwave electronics. The analysis was also dependent on the use of highly developed theoretical concepts such as k•p perturbation and symmetrical considerations of space group.

In the latter half of the decade more quantitative studies were developed both theoretically and experimentally. The band calculation was very much developed by the introduction of the concept of pseudo potentials by J. C. Phillips and it opened a way for comparative studies of band structures in various semiconductors such as Si, III-V compounds and even an alloy of Ge and Si.

The effective mass theory for the dynamics of carriers developed by W. Kohn and J. M. Luttinger played a very important role in a phenomenological analysis of various experimental data. The successful analysis of the electronic structure of the ground and excited states of donors and acceptors, and a description of the warping in hole bands near the Γ-point as well as the Landau levels of these degenerate bands were examples of such analysis.

As to the transport theories, a precise analysis was performed for the transport coefficients of carriers in many valley semiconductors such as germanium by C. Herring in 1955, which was based on the Boltzmann's transport equation. A general quantum mechanical expression for transport coefficients was developed by R. Kubo in the latter half of the decade. P. W. Anderson's paper on the absence of diffusion in so-called impurity conduction phenomena was presented in 1958. The phenomena were first found by C. S. Hung and J. R. Gliesmann in 1951. These papers are now playing a basic role at present in the recent researches on quantum transport phenomena and on the effect of localization of carriers in semiconductors.

There was much progress in experimental investigations too. To show some examples, I have picked up the following reports presented at the 4th I.C.P.S. at Rochester.

The magneto-optical studies by B. Lax et.al. by using a very big magnet established at M.I.T. inaugurated a series of precise measurements of Landau level structures and exciton levels of germanium and various semiconductors in the following decade. The spectroscopic studies of optical absorption of donors by H. Y. Fan and his collaborators in Purdue, the observation of phonon drag effects in germanium by T. H. Geball, C. Herring and J. E. Kunzler, and inelastic neutron scattering experiment for the determination of phonon dispersion relation of germanium by H. N. Brockhause and a mapping of the trapped electron density near a donor center in Si by the method of ENDOR done by G. Fehr which successfully determined the k points of the valleys of the conduction band in Si, were examples of remarkable reports made at the Rochester Conference in 1958.

Thus I suppose that the reader can easily understand that solid state physics had indeed reached a stage of "exact science" in terms of germanium single crystals and that the Rochester Conference was really a remarkable mile stone in the history of modern semiconductor physics. As pointed out by Dr. Frova in his closing address at the Roma Conference in 1976, the activity of germanium studies decayed gradually after the Rochester Conference and the interests shifted to other materials such as Si and III-V compounds in the next decade, although the first report on the semiconducting properties in III-V compounds such as InSb were given by H. Welker in 1951.

One of the distinguished characteristics of the semiconductor research in the first decade was the very good mutual interactions between basic and technological fields. The scientists and engineers shared their common interests in various phenomena of semiconductors, and they gathered and discussed with one another frequently in the same sessions at the same conferences such as I.C.P.S.

The dislocation in Ge and Si were a subject of common interest for both basic physics of dislocations and material technology for good transistors. In the 3rd I.C.P.S. at Garmischpartenkilchen E. Billig was invited to give a talk on the role of dislocations in the preparation of large single crystals of germanium for device usage. Direct observation of dislocations in highly perfect germanium single crystals by an etching technique and in Si by using a infra red transmission microscope and decoration techniques gave a vivid evidence for the model of small angle grain boundary, for the stress field around a dislocation line and for the model of Frank-Read multiplication mechanism of dislocations.

The recombination life time of minority carriers was another example of the phenomena of common interest. The sessions named "recombination" were frequently arranged in I.C.P.S. during the first decade. How to determine the cross section of the radiationless recombination experimentally and how to interpret its value and temperature dependence theoretically were the basic subjects to be studied from both academic and technological points of view.

As to the semiconductor devices, p-n junction and junction transistors were brought to the stage of quality controlled mass production during this decade. At the same time, the I-V characteristics of very good p-n junctions revealed the first experimental verification of the theoretically predicted typical exp(eV/kT)-1 dependence showing universal rectifying character independent of junction properties.

In the latter half of this decade, break down phenomena in the backward I-V characteristics of p-n junction were attractive both from the physical and technological points of view. The attention was paid in order to produce high quality junction rectifiers because of an industrial need. However, the attention was also paid to show a definite experimental evidence of the so-called "Zener tunnelling" of carriers between valence and conduction bands because of basic interest.

Thus one might say that two sectors, inductrial and academic were in their honeymoon in the first decade and as a result Esaki's tunnel diode was born in 1957 in which clear evidence of tunnelling was given in observed reverse polarity to those in usual diode and of negative resistance in the usual forward direction was first demonstrated.

What were the Japanese activities on semiconductor physics in the first decade? Simply speaking, this decade was a period of recovery and reconstruction from the very big damages of World War II. Although there were many barriers in communication, the news of the transistor invention reached Japan under occupation rather quickly, and in 1949 the first two papers, the one on transistor action (theoretical) and the other on rectifying action in germanium were read at the meeting of the Physical Society of Japan (P.S.J.).

During the war, some research activities were developed on Si point contact rectifiers as detectors of RADOR wave. However, there were no activities on germanium at all. The activities on rectifiers of Si point contact and Se planar contact as well as on the material researches for oxcide cathodes and phosphorus were developed mainly in laboratories of some electronics companies. Many scientists and engineers who were working in these fields were a source of specialists for the development of modern semiconductor physics and technologies in Japan. Their number grew rapidly in the latter half of the decade. Another source was groups of scientists studying color centers in universities including theoretical physicists.

It is interesting to know that the semiconductor physics in Japan had already been in close contact with electronic industries since its beginning, and had been rather free from the influence of strong traditional schools in academic circles although there were minor experimental groups to study rectification mechanism in some universities. It shows a distinguished contrast to the field of magnetism in Japan. The reader is no doubt acquainted with the fact that activities in this field have been strong traditionally in Japanese universities since the early days of the 20th century. The names of Profs. K. Honda and S. Kaya have been well known internationally in the field of magnetism since before the war.

In 1950, the year of the first I.C.P.S., Japan was still under occupation and no Japanese participant was found at the Reading Conference in 1950.

After the peace treaty in San Fransisco in 1951, an "International Conference of Theoretical Physics" was held in Kyoto in 1953, and Prof. N. F. Mott was the chairman. It was the first large scale international conference of physics in Japan, so that the conference was a monumental one to announce reemergence of Japanese physicists in the international academic circle, and it gave great stimuli to the research activities at that time as an ignition effect for the post war developments. In the field of solid state theory, many distinguished physicists came from abroad to attend the conference, such as N. F. Mott, F. Seitz, J. C. Slater, J. H. Van Vleck, R. E. Peierls, L. Néel, N. Bloembergen, H. Fröhlich, P. W. Anderson and J. Bardeen. Dr. Bardeen's visits to some laboratories after the conference were very fruitful for the Japanese semiconductor physicists working in these laboratories who had just started germanium research.

In the 2nd I.C.P.S. at Amsterdam in 1954, the first paper presented at the I.C.P.S. from Japan on the theoretical investigation of the anisotropic galvanomagnetic effect in germanium by M. Shibuya was read by M. Hatoyama. Only two participants, Dr. Hatoyama and Prof. Muto, attended the conference from Japan.

Although the first theoretical paper was presented at the 2nd I.C.P.S. in 1954, the experimental studies of germanium in Japan needed some inclubation times to produce original papers on germanium because of various barriers in experimental conditions.

One of the barriers was the sample quality and it was mostly overcome by the end of the first half of this decade. The year 1954 was the year during which industrial production of germanium transistors began. The first commercial transistor radios were put on the market in 1955 by Totsuko Co. (the predecessor of SONY). The great efforts to obtain good single crystals and p-n junctions were made in some industries and governmental or semi-governmental institutes. Reflecting these activities many symposia and special sessions were organized during the P.S.J. meetings. The segregation effect of impurities, the dislocation properties, and electron hole recombination were the typical topics discussed at these meetings. At the 3rd I.C.P.S. at Garmischpartenkilchen in 1956, H. Kawamura et.al. presented a paper on the trapping and the recombination life time caused by dislocations in alkali halides and in germanium single crystals. The experiments were done by use of germanium samples grown in domestic companies.

At the annual meeting of P.S.J. in 1957 one could listen to a report by L. Esaki and Y. Kurose on "The internal field emission of Ge p-n junction" and it was followed by two papers on "Studies of Ge p-n junction showing negative resistance due to Zener emission II and III" by L. Esaki et.al. read at the spring meeting of P.S.J. in 1958. These were the reports presented at P.S.J. meetings on Esaki's famous work on tunnel diodes for which he was awarded the Nobel Prize in physics in

1973. I am quite sure that these results were brought about by Dr. Esaki's ingenious talent for physics. However it is also true that the standard of sample technologies on germanium in Japan had already reached a high stage by then.

The latter half of the decade was just the starting period of the rapid growth of industrial and economical activities in Japan. The newly developed semiconductor industry was one of the young champions to play an important role in the growth. Being supported by the recovery and reconstruction of the national economy, the barrier in experimental facilities were also gradually removed and the researches on the bulk electronic properties of germanium single crystals were developed not only in the industrial laboratories but also in some government or semi-government institutions and universities. In 1950, we had no low temperature facilities of liquid helium in Japan, and as a consequence one could not determine the binding energy of carriers at a donor center in germanium by measuring the temperature dependence of carrier concentrations through Hall coefficient. However, in the latter half of the decade the low temperature facility was first constructed at the "Institute for Study of Iron and other Metals" of Tohoku University. The institute also started a high magnetic field project in this period in the leadership of Prof. T. Fukuroi. In 1957, the government decided to establish a new institute, now called "Institute for Solid State Physics" (I.S.S.P.) which was attached to the University of Tokyo but opened to all solid state scientists to use various facilities in the institute. It was also expected to play a role of as a center of activities in various fields of solid state physics. The first director of the insitute was Prof. S. Kaya and Profs. H. Kawamura and Y. Toyozawa were first invited to start research activities in the semiconductor section of the institute.

Thus, at the 4th I.C.P.S. at Rochester in 1958, experimental reports on bulk semiconductors such as the observation of anisotropic conduction of hot electrons in germanium by W. Sasaki et.al. and the measurement of the thermal conductivity of InSb and PbTe by Y. Kanai and R. Nii were presented from Japan.

The Phenomena of impurity conduction were attractive to some Japanese semiconductor physicists since its first discovery in 1951. T. Kasuya and S. Koide discussed the theory of hopping mechanism of impurity conduction. W. Sasaki et.al. observed negative magnetoresistance of n-type germanium at low temperature, and Y. Toyozawa presented a model to explain the negative magnetoresistance by means of localized spin supposed to be caused by the random site distribution and correlation between carriers. These works were reported at the Prague Conference of I.C.P.S. in 1960.

As shown in the curves of Fig. 1, the start of the rapid rise of the curves is found in the latter half of this decade. The reader can also find a peak structure in the curves of Fig. 2(a) and Fig. 2(c). Through the brief sketch just presented the reader should be able to identify the origin of the structure observed in these curves.

I think that I have used too many pages on a sketch of the "Dawn of Modern Semiconductor Physics". In those days, I was a young physicist and I had just started my semiconductor research at Toshiba Electric Co. after finishing my graduate work on the electronic structure of colour centers in Alkali Halides. I was very much impressed by many distinguished papers on semiconductor physics published mainly in the Physical Review by the physicists in Bell Telephone Laboratory. I reached to the conclusion that the most important and urgent research was to produce good single crystals of germanium in Japan. Thus I joined a research group for crystal growth and studied the properties of dislocations in germanium. I could verify the law of elastic interaction between two dislocation lines to analize the configuration of paired or multiple structure of etch pits along a small angle grain boundary in germanium. My memories of those younger days are still very fresh so that I have used too many pages unwarily for my "Dawn of Modern Semiconductor Physics".

Presumably many semiconductor physicists in a rather older generation who once played his major roles in those young and animated days of semiconductor physics may recall their younger days with some kind of emotional sentiment.

3. Extension and Sophistication

The heading for the second decade (1960-1969), "Extension and Sophistication", was chosen from the main theme of the opening address by Prof. B. Lax at the Cambridge Conference (the 10th I.C.P.S.) in 1970.

What is the extension in semiconductor materials to be studied in research laboratories and produced in industrial factories? First of all, I will mention that single crystals of silicon became a main actor in semiconductor physics and technologies in this decade. In the previous decade, it was said that there were barriers for the use of Si for junction transistors instead of Ge, although Si has many favourable properties for electronic devices such as its thermal stabilities, a suitable value of its forbideen energy gap and its easy availability. One of the barriers was found in the difficulty to obtain high purity in the crystal of Si free from oxygen impurities. This barrier was removed by the development of the so-called "floating zone method" in zone refining processes by which a sample was kept free from contamination by impurities from the wall of the crucible. This method was applied to produce Si crystals in the first half of the decade. It was interesting to find that this disadvantage of Si being chemically reactive with oxygen turned out to be an advantage for semiconductor devices in the latter half of the decade for the production of I.C. (integrated circuit) which had become a new stream in semiconductor device technologies instead of Ge p-n junctions and junction transistors.

Although there was a pioneering work on the III-V compound semiconductors by H. Welker in 1951, considerable progress in the research of these compound semiconductors was made in this decade. The progress was partly a result of gradually improved material technologies which enabled to controll the stoichiometry of these samples.

The conduction band of III-V semiconductors has second minima in electronic energies at the symmetrical k points aside from the first minimum point at k = 0 in the first Brillouin zone. The difference in energies between the first and second minima is so small that one can expect to realize a considerable population of carriers in these valleys of second minima having a larger effective mass than that of the first minimum by heating the conduction electrons by applying high electric field. The Gunn effect observed in non-linear I-V characteristics of n-type GaAs was a typical example caused by the instability due to this distribution of hot carriers. The discovery of this effect by J. B. Gunn in 1963, led GaAs to become a basic material for the device called the "Gunn diode" — solid state alternatives of the

"Krystron".

The III-V semiconductors became attractive as materials of opto-electric devices in the latter half of the decade with the development of emission diodes and junction lasers. The laser action in semiconductors was first observed during the period 1962-1963 in the GaAs p-n junction at 77°K as pulsed oscillation. New technologies were rapidly developed in this second decade to realize stable oscillation, and finally the diode began to play an important role in modern opto-electric devices.

As shown in Fig. 2, not only the III-V semiconductors but also many kinds of semiconductors such as II-VI compounds (CdS and PbTe etc.) were studied extensively during this decade. The studies of elemental semiconductors and semimetals such as Te, Bi and graphite were also made in this second decade, which showed various interesting properties characteristic of each material. These extensions of materials to be studied were strongly supported by the development of many theoretical and experimental methods in solid state research which were achieved in the first and in this second decades. In the 8th I.C.P.S. (Kyoto 1966) the semiconducting properties of magnetic compounds, MnP and MnS, were reported in connection with their magnetic properties. A possibility to realize the superconducting state in semiconductors was studied theoretically and was also studied experimentally for the degenerate semiconductors of GeTe and $SrTiO_3$.

Now, I will consider the "extension and sophistication" in the scope of methods used in the investigation of semiconductor physics. It was quite natural that the phenomena of electrical conduction were the first properties to be investigated at the dawn of modern semiconductor physics. Indeed, various important contributions were made on the studies of semiconducting properties of Ge during the previous decade. These studies gave a kind of standard form of investigation on the transport properties in various semiconductors, semimetals and even in metals.

However, the current of research in semiconductor physics was gradually changing in this second decade. As pointed out by Dr. Frova in his closing address of the 13th I.C.P.S. (Rome 1976), the optical or spectroscopic studies became one of the very active streams in this decade. The spectroscopic studies can approach directly to the details of macroscopic properties. As a result the tendency towards sophistication was inevitable for the research of this decade.

Following the optical absorption measurements of band gaps and of donor and acceptor states in the previous decade, measurements were extended to observe continuous spectra of reflection caused by transitions between valence and higher conduction bands. For germanium, H. R. Philipp and E. A. Taft had already observed optical constants in the region of 1 to 10 eV in 1959. A similar investigation was soon applied to Si and Ge-Si alloys in this second decade. The studies of optical reflection spectra in III-V compounds were also developed, and reported mainly at the 7th and the 8th I.C.P.S. held in Paris 1964 and in Kyoto 1966. The method of so-called modulation spectroscopy such as piezo-optical, electro-optical and thermo-optical as well as well established magneto-optical measurements helped to obtain very precise data on the structures of reflection spectra. The theoretical investigations on the microscopic descriptions of dielectric functions for optical constants, of the effect of the lack of inversion symmetry in III-V semiconductors on the band structure and of the detailed calculations of the optical density of states and its van-Hove singularities were the back ground, which supported the interpretation and assignment of these detailed optical data.

Exciton in semiconductors such as Cu_2O and CdS was also a very attractive subject for optical studies in this second decade. It was opened by the report of precise spectroscopic measurements given by E. F. Gross and S. Nikitine at the Prague Conference 1960, followed by the application of the magneto-optical technique to observe rotary dispersion and Zeeman effects. At the Kyoto Conference of 1966, an interesting theoretical aspect of the coexistence of local and band structures in the absorption spectra of exciton in semiconductor alloys was presented by Y. Toyozawa, M. Inoue, T. Inui, M. Okazaki and H. Hanamura.

In the 9th I.C.P.S. in Moscow 1968, many important contributions concerned with semiconductor lasers were reported. Thus a new stage was opened for the optical studies of semiconductors. A. A. Rogachev produced a very high density of excitons in Ge by laser irradiation. It indeed ignited the study of high density excitons which produced many fruitful physics in the next decade. The observation of light scattering from mobile carriers was also reported in the Moscow Conference 1968 by P. A. Wolf.

The spectroscopic studies on the donor and acceptor states in III-V semiconductors were also very much stimulated by the development of emission diodes and junction lasers. I was very much impressed by the report on the so-called pair spectra of electron hole recombination between distant pairs of a neutral donor and a neutral acceptor by which a determination was made on the location of donor and acceptor atoms in the crystal. This report in I.C.P.S. was given at the Paris Conference 1964 and the main contributors on this subject during the decade were D. G. Thomas, E. M. Gresshenzone and J. J. Hopfield.

As to the theoretical studies, the band calculation improved its reliability quantitatively due to the progress of computer. At the same time the concept of quasi particle or of elementary excitation for many body interaction was applied to the semiconductors. The concepts of plasmon, of magneto-plasmon, and of polariton were introduced in this decade in addition to those of exciton and polaron which had already been known in the previous decade.

These quasi particles were studied experimentally by observing the propagation of electromagnetic waves in semiconductors under various configurations of polarization and static magnetic fields. Helicon waves and Alfvén waves in degenerate semiconductors and semimetals were studied extensively as well as magneto-plasma cyclotron resonances. Many interesting results were reported at the satellite conference on the waves in solid state plasma of the 7th I.C.P.S. in Paris 1964. The interaction of helicons with other excitations in solids reported by P. Aigrain, infra-red reflectivity of Bi in the quantum limit reported by L. C. Hebel were examples of these results.

Transport studies were developed in two directions: one was the progress in non-linear conduction and instabilities caused by high electric field. The Gunn effect, already mentioned, and phonon amplification by hot carriers in CdS were examples of the investigation in this direction. This direction was an extension of hot electron studies which were started by the pioneering work of Ryder and Schockley in 1951 and developed in detail during previous decade and the first half of this decade. Another direction was to find or to clearify the quantum effect in low temperature transport phenomena in semiconductors. The experimental studies on oscillatory magneto-conductance in Si surfaces was reported by A. B. Fowler, F. F. Fang, W. E. Howard and P. J. Stiles at the 8th I.C.P.S. in Kyoto 1966. It was the first paper for the investigation of quantum transport phenomena in two dimensional carriers under high magnetic field. It was followed by many interesting investigations developed in the next decade. Studies of impurity conduction made progress by the observation of its frequency dependence over a wide frequency range. At the Prage Conference 1960, M. Pollak and T. H. Geball reported on the frequency dependence of dielectric loss of Si in the low temperature range caused by a fluctuation of the dipoles consisting of carriers in impurity centers and charged minority centers. H. Y. Fan and S. Tanaka first observed direct resonance absorption of a microwave quantum by an ionized molecular pair of donor or acceptor centers. Detailed reports were given at the 7th and 8th I.C.P.S. in Paris 1964 and in Kyoto 1966.

As given in this brief sketch, "extension and sophistication", the physics of semiconductors in this decade became one of the peaks in the mountain range of solid state science. However it is also true that a tendency toward sophistication in research activities meant getting involved with too many details and somewhat moving away fresh and naive new ideas. Indeed, the theoretical development of many body interactions in particular of superconductivity or various new findings in the field of biophysics in this decade were seen to be much more attractive in freshness compared to the "extension and sophistication" in semiconductor physics. I will refer to the following words given by Dr. Aigrain in his concluding talk at the 7th I.C.P.S. in Paris 1964.

"Ten years ago, the field of semiconductor was one which was very unsophisticated, very exciting because it was unsophisticated; it was somewhat like Marilyn Monroe in her first movies: The field is now much more like some of the latest movies with Marilyn Monroe. It has become very much more sophisticated; it does not prove it has less interesting."

The semiconductor technology too, became a star of modern industries with newly developed devices such as Si-MOSFET, junction lasers and the Gunn diodes, so that the term "solid state electronics" became a household word and began appearing even in TV commercials. However, we should not omit the start of a very important and new technoligical development, namely that of integrated circuit. It was an embryo of the L.S.I. and V.L.S.I. recently developed to meet ever escalating need for larger and faster computers.

Indeed, a start for "micro-electronics" was initiated by the invention of MOSFET (Metal-Oxcide-Semiconductor-Field-Effect-Transistor) based on conductivity modulation by gate voltages. The idea of field effect transistors (FET) had been proposed prior to the discovery of the usual transistor action caused by the injection of minority carriers. The improvement of technology to obtain a good interface between Si and SiO_2 made it possible to realize the original idea of field effect transistors. The patent for MOS was granted in 1960 in U.S.A.. In 1964, the integrated circuit (I.C.) of MOS was produced for the first time and in 1968 200-bit static shift-registers (nearly 1400 pieces) were developed. It was just an opening of the so-called "era of L.S.I.".

Sophistication in academic research and the start of a new development in industrial research brought about the end of a honeymoon enjoyed by the both academic and industrial groups of researchers. From then on, solid state physicists and device physicists and engineers began to walk their own separate ways and to organize their own conferences separately. That is one of the characteristics of semiconductor research at the end of the second decade of "extension and sophistication" in semiconductor physics.

What was Japanese semiconductor physics like in the decade? Generally speaking its characteristic can also be summarized by "extension and sophistication" as in the case of international trends. The international exchange of physicists and of scientific information increased very rapidly in this decade. Many young scientists were invited to go abroad and to stay as research fellows in foreign universities and research organizations. The supports given by the Fulbright Committee in U.S.A., by the British Council, by the Humbolt Foundation and by many other foreign foundations helped to accelerate it very much. The number of Japanese participants in international conferences was also increased. It was our great pleasure that the 8th I.C.P.S. was held in 1966 in Japan, the first Kyoto Conference of I.C.P.S. Prof. T. Muto and Dr. M. Hatoyama were the chairman and the general secretary of that conference, respectively. Thus the information gap existing in the previous decade had almost vanished. This was one of the reasons that the characteristic of Japanese semiconductor physics in this decade was not so different from that of the international one.

As shown in Fig. 1, a rapid increase in the number of published papers on semiconductor physics in J.P.S.J. is observed in the first half of the decade and then it seems to be saturated in the second half of the decade. It was part of effects caused by the rapid increase in the population of semiconductor physicists as a result of educational effort in many universities in Japan. The number of department of applied physics and electronics increased rapidly on the engineering side while an increase in the number of solid state physicists was also seen in the traditional department of physics on the pure scientific side. As a result, the number of members of "The Society of Applied Physics of Japan" expanded very much and its domestic annual meeting has been held separately from that of the P.S.J. since 1965. At the same time, both societies jointly began publishing a new journal "Japanese Journal of Applied Physics" (J.J.A.P.), for papers having rather device oriented interests. This is the reason for the saturation in the number of papers on semiconductor physics in J.P.S.J., because device oriented papers were submitted to the J.J.A.P.

The activities in semiconductor research soon became a main stream of the activities of the Society of Applied Physics in Japan. The number of papers on semiconductors read in the meetings of the Society in a year was about 150 in 1965. It reached nearly 300 in 1970 and close to 650 in 1980. The papers on semiconductor physics read in the meetings of P.S.J. in a year fluctuated between around 200 and 300 since 1965. These numbers suggest that the research on semiconductors were classified into two streams, the one is physics oriented and the other is device oriented. These two streams were seen to walk gradually in their own separate ways. I think that the rise of activities was found not only in the quantitative side as just mentioned but also in the qualitative side as well. The continuous increase of accepted papers submitted to every I.C.P.S. shown in Fig. 1 seems to suggest that the quality of the activities of Japanese semiconductor physics in this decade was well appreciated in the international academic community.

In order to end this section, I will describe, in the following, a brief sketch of each research of Japanese semiconductor physics in this second decade.

In the first half of the decade some sophisticated and important contributions were made on the fundamental bulk properties of Ge and Si. H. Miyazawa measured the anisotropy in galvano-magnetic properties of holes in Ge as a function of electron temperature both in the normal and in the hot electron conditions. In cooperations with young theoretical physicists, H. Maeda and K. Suzuki, he succeeded in elucidating the difference in the nature of warping between the hole bands of Ge and Si. As to the quantum effect in Landau levels of holes in Si and to the effect of external stress on the cyclotron resonance of electrons and holes in Si, J. C. Hensel in B.T.L. performed very comprehensive and beautiful experimental studies which gave us a precise knowledge of the interband mixing caused by the presence of spin orbit coupling and of external magnetic and stress fields. In Japan, M. Okazaki first showed keen theoretical interests in this quantum effect and A. Hasegawa, H. Kamimura and K. Suzuki were invited to stay at B.T.L. to analyze his work from a theoretical point of view. These activities for the band structures of Ge and Si were reported in every I.C.P.S. from Prage 1960 to Kyoto 1966 by H. Miyazawa and J. C. Hensel. The quantum effect of cyclotron resonance in holes of Si was finally reported at the 10th I.C.P.S. in Cambridge in 1970. This work was a typical example of "sophistication". Thus all phenomenological parameters to describe Landau levels of valence bands in Ge and Si introduced theoretically by J. M. Luttinger were determined by experiments done by J. C. Hensel.

To observe cyclotron resonances in Ge, physicists in Japan faced various difficulties in experimental conditions because of a lack of general background and of facilities in Japan at that time. Then the first experimental success was obtained by Prof. H. Kawamura and his collaborators after I.S.S.P., the newly established institute, started to function. The first paper was presented at the 6th I.C.P.S. in Exeter 1962 by H. Kawamura, I. Imai and K. Tomishima on the cyclotron resonance of hot electrons in Ge. Prof. Kawamura did not intend to discuss the effective masses from the resonance fields, but rather he observed the line shape of the cyclotron resonance to study the scattering mechanisms of carriers such as the phonon scattering rate in the low temperature range of 4°K, electron-hole scattering, the rate of intervalley scattering etc. These results were reported in every I.C.P.S. from that of Exeter, 1962 to that of Kyoto, 1966. The research in this direction was followed by E. Otsuka and the cyclotron resonance study of electron scattering by deep level acceptors in Ge was reported by E. Otsuka, K. Murase and K. Takesawa at the 9th I.C.P.S. in Moscow 1968.

As I have already mentioned, the cyclotron motion of carriers couples with the plasma wave or magneto-plasma wave when the sample is degenerate semiconductors or semimetals having a rather high concentration of carriers compared to the usual semiconductors. The degenerate sample of PbTe was studied by Y. Kanai for helicon wave and by R. Nii and A. Kobayashi for magneto-plasma cyclotron resonance. H. Numata and Y. Uemura analyzed their data theoretically and explained that the detailed structures in the resonance pattern were all interpreted by the dielectric anomalies, hybrid resonances and tilted orbit cyclotron resonances for both n and p type samples. The equi-energy surfaces near the bottom of the conduction band and near the top of the valence band in PbTe were determined to be similar to the structures of the conduction band in Ge. These results for PbTe were all reported at the satellite conference of the 7th I.C.P.S. in Paris, 1964. H. Kawamura extended his cyclotron resonance studies to the magneto-plasma wave of semimetals such as Bi after his moving from I.S.S.P. to Osaka University. At the 9th I.C.P.S. in Moscow 1968, H. Kawamura, S. Takano and O. Shimomura reported on the magneto-plasma waves in Bi under extreme quantum conditions. Further, he observed and analyzed the nonlocal effect in the magneto-plasma cyclotron resonance in Bi with J. Nakahara.

As to the studies of semimetals in Japan, S. Tanuma studied the diamagnetism of Bi, BiSb, As and graphite. S. Mase observed giant quantum oscillations in the magneto acoustic effect in Bi. K. Sugihara, M. Tsukada and Y. Uemura studied the electronic properties of graphite theoretically and its experimental investigation were done by T. Tsuzuku and S. Tanuma.

Studies of optical properties of semiconductors developed very much in this decade. Although many investigations had been reported on magneto-optics in the previous decade by B. Lax et.al., no contribution was made from Japan in the previous decade. At the 7th I.C.P.S. in Paris 1964, A. Misu, K. Aoyagi and G. Kuwabara reported on the absorption edge of CdS in strong magnetic field produced by condenser discharge. At the 8th I.C.P.S. in Kyoto, 1966, magneto-optical studies were extended to layer type semiconductors such as GaSe. The result was reported by many authors contributed to the investigation. The band structure of GaSe was also studied phenomenologically by H. Kamimura and K. Nakao. Y. Nishina studied his magneto-optical studies on excitons and Y. Hamakawa started the electro-optical studies after their return from their studies in U.S.A. Their results obtained in Japan were reported at the 8th and the 9th I.C.P.S. in Kyoto 1966 and Moscow 1968.

Y. Yoshinaga was a pioneer in Japan to start the infra-red optical studies of semiconductors and reports on local vibrational frequency of Li and Al in ZnSe or in ZnTe were read at I.C.P.S. in Kyoto 1966 and Moscow 1968.

Theoretical studies on the optical properties of semiconductors were stimulated by the distinguished contributions by Y. Toyozawa on the excitons. After developing the line shape theory done in the previous decade, he introduced various new concepts in this field. The distinction between the so-called isolated and amalgamated types of exciton in alloy semiconductors was presented at the 8th I.C.P.S. in Kyoto as already mentioned. He also developed the studies of effects of the exchange interaction on exciton absorption with Y. Onodera. Reflecting his contributions made in this decade, he was asked to give an invited plenary talk on "the new aspects of excitons" at the 10th I.C.P.S. in Cambridge 1970. As to the theory of impurity states, A. Morita and H. Nara reported on the chemical shift on shallow donor states in Si, and H. Kamimura reported on the states of Cu impurity in II-VI compounds which was investigated with R. E. Dielz when he stayed at B.T.L.

Finally, I will sketch the transport studies done in Japan during this decade. In 1960, W. Sasaki, M. Yamanouchi and M. Hatoyama reported on their observation of negative magnetoresistance in the so-called impurity band conduction phenomena at the I.C.P.S. in Prague. In 1962, Y. Toyozawa proposed a model of localized spin interacting with conduction electrons in order to understand those phenomena. Since then, the problem was investigated intensively for various concentration ranges of impurities. The results were reported by W. Sasaki at the 8th I.C.P.S. in Kyoto. A new trend in the field was brought by the introduction of microwave studies. One is the observation of the E.S.R. line shape in various concentration ranges reflecting the motional narrowing due to impurity conduction. K. Morigaki, S. Maekawa and H. Kodera were the contributors in this field. Their reports were found in I.C.P.S. in Kyoto 1966 and in Moscow 1968. The other is the direct observation of microwave resonance absorption caused by the ionized pair of donors or acceptors in Si. The study was started by H. Y. Fan and S. Tanaka when S. Tanaka was staying in Purdue University. He developed the investigation very intensively after his coming back to Tokyo with his young collaborators. As I have already mentioned the reports were found in I.C.P.S. of Paris 1964 and of Kyoto 1966. The theoretical problems involving impurity conduction phenomena were randomness, correlation, and their interplay. T. Matsubara and Y. Toyozawa performed a pioneering work on the treatment of randomness reported by Y. Toyozawa at I.C.P.S. in Prague 1960. The band type or localized type characters of the exciton in alloy semiconductor discussed by Y. Toyozawa, already mentioned was also a contribution as a natural extension of his original idea. M. Tsukada and Y. Uemura studied the effect of the stacking disorder distributed randomly in the layer structure of graphite on the modification of Fermi surface. A report was made at the I.C.P.S. in Moscow 1968. Following a theoretical report on the Hall effect by R. Kubo read at I.C.P.S. in Paris 1968, quantum transport phenomena attracted much attention. S. Kawaji and Y. Kawaguchi observed negative magnetoresistance phenomena in the conduction along InAs surfaces. It seemed to suggest that the effect caused by the localized spin due to an un-paired orbit of surface atoms. A report was given at I.C.P.S. in Moscow 1968. This was a pioneering work in Japan on the studies of transport phenomena at the interface which developed rapidly in the next decade in Japan reflecting the progress of MOSFET and I.C. technologies.

Hot electron was another field developed very much during the decade. J. Yamashita was a pioneer in the theoretical research on hot electrons in Japan. He found an analytical solution of Boltzmann's equation keeping the non-linear character of external field. In this decade he discussed the role of electron-phonon interaction in the hot electron condition and in instabilities when drift velocity of carriers reaches the sound velocity. The results were reported at I.C.P.S. in Paris 1964 and in Kyoto 1966 together with K. Nakamura. T. Kurosawa introduced the Monte Carlo method in the hot electron theory which was reported at I.C.P.S. in Kyoto 1966. He also discussed the hot electron problem in high magnetic field with E. Yamada which was stimulated by an experiment on InSb done by K. Komatsubara, N. Kotera and E. Yamada. These studies were reported at I.C.P.S. in Kyoto 1966 and in Moscow 1968.

As to the experimental studies on instabilities there was no scientific contribution in Japan on the Gunn effect, although device oriented researches were interested in it. On the contrary, instabilities caused by the coupling with phonons attracted considerable attention. In the I.C.P.S. in Kyoto 1966, a paper on current saturation and instabilities in CdS was reported by A. Ishida, C. Hamaguchi and Y. Inuishi and on the behavior of phonons in CdS was reported by E. Maruyama and Y. Murayama. T. Ishiguro and T. Tanaka discussed the ultrasonic amplification and non-ohmic behavior in Te, and T. Arizumi, M. Umeno and S. Iwai studied the build-up of phonons in Ge experimentally.

So far I have given a brief sketch of semiconductor physics in Japan during this decade. Comparing it with the previous decade, the reader can find that the research activity reached a standard such that Japanese physicists could join the international academic circles in various fields of semiconductor physics. The information gap and the time lag of response to new trends almost vanished during the decade. However, frankly speaking, it seemed to need a little bit more incubation time to produce truely original stream of semiconductor physics characteristic of Japan. This might be an inevitable tendency of the decade shown by a rapid growth of international exchange and by various studies opened by young physicists who had just come back from abroad during this decade. Thus "extension and sophistication" is also a suitable phrase for summarizing the Japanese activities in this decade.

4. Sign of Transformations (1970-1980)

As the heading of the third decade (1970-1980), I proposed "Sign of Transformations" in the closing address of the 15th I.C.P.S. in Kyoto, 1980. Strictly speaking the decade should end in 1979. However, I will include the results of this Kyoto conference for convenience.

Of course, one can recognize a steady and continual development in various fields of semiconductor physics. However, the birth and rapid growth of new fields during this decade have been so remarkable that trends in research seem to indicate a sign of transformations. It is quite certain that transformation has partly been brought about by technological impacts. The progress of microelectronics brought many new technologies to semiconductor industry. The needs for surface studies from the microscopic view point as well as those for the space charge layer of interface came about by the process engineering of L.S.I. and V.L.S.I. The needs for investigation of amorphous substances were also arose by the engineering needs for semiconductor solar cells and for the display technologies. These are some examples of the impact. Thus the difference between academic and industrial research which had been increasing in the previous decade again began to decrease because of shared interests, although the size of the numbers of researchers in both sectors made it impossible to organize a single large scale conference as had been done in the period of the dawn of modern semiconductor physics.

To remedy this situation and to build bridges between these sectors, a special session has been organized in I.C.P.S. since the 13th I.C.P.S. in Rome 1976. The speakers and their subjects in this first special session were: "Highlights in semiconductor device development" by L. Esaki, "Hot electron devices - from physics to applications", by C. Hilsum, "Fundamental physical limits to large scale integration" by J. T. Wallmark, "Recent advances in heterojunction lasers" by H. Kessel and "Semiconductors in solar energy conversion" by B. O. Seraphin. From these subjects you can easily understand the origin of technological impacts just mentioned. At the 14th I.C.P.S. in Edinburgh 1978, two sessions were devoted for this purpose: one was for the interaction of laser radiation with semiconductors and the other for the introduction and characterization of defects in semiconductors". Both sessions consisted of four invited talks. From Japan, E. Otsuka was invited to talk on "Far infra-red spectroscopy" in the former session.

At the 15th I.C.P.S. in Kyoto 1980, reflecting the recent interests in material technology for L.S.I. and for opto-electric devices, two challenging talks to semiconductor physicists were presented by Japanese physicists working in industries to stimulate various comments and discussions. The topics were: "The role of oxygen and carbon in Si crystals in V.L.S.I. technology" by S. Kishino and "Degradation in III-V opto-electric devices" by I. Hayashi.

I will remind you, here, that the impacts which were a result of the transformation in semiconductor device technology, caused a sign of transformation in very basic theoretical studies. Indeed, this is a traditional characteristics of semiconductor research. As an example, I will refer to the investigations on two dimensional carriers in the space charge layer of semiconductor interface developed rapidly in this decade. There is no need to say that the study was closely related to the development of MOSFET technology. The improvement in controlling the so-called surface states caused by the misfit of the interface was the key not only to produce good MOSFETs but also to investigate various interesting phenomena of the physics of two dimensional carriers in the space charge layer of the interface. I have already mentioned in §3, the first distinct experimental evidence of the existence of quantized states for carriers in the MOS inversion layer was reported by A. B. Fowler et.al. at the 8th I.C.P.S. in Kyoto 1966. The detailed theory based on Hartree's self-consistent method was reported by F. Stern at the 10th I.C.P.S. in Cambridge Boston 1970. Since then, the field developed very rapidly and the following plenary invited talks were given at the I.C.P.S. in Warsow 1972, Stuttgart 1974, Rome 1976 and Edinburgh 1978. They were: "Quantum effects in a semiconductor surface layer" by D. C. Tsui, "Quantum galvanomagnetic phenomena in MOS inversion layers" by Y. Uemura, "Remarks on the spectroscopy of electronic states in surface space charge layers" by F. Koch and "Transport in quasi-two-dimensional space charge layers" by P. J. Stiles, respectively. During the decade, the subband structures, their Landau levels, the effect of many body interaction between carriers in these states, the spin and valley splitting of Landau levels, the exchange enhancement of the splitting, and the magneto-conductivity tensor of static and dynamical conditions under the extreme quantum limit have been thoroughly investigated both theoretically and experimentally. T. Ando was a distinguished contributor to the theoretical developments in this field. The results were summarized in his invited talk at the 15th I.C.P.S. in Kyoto 1980.

At this Kyoto Conference of 1980, two very remarkable experimental results were reported. One was a report by Y. Kawaguchi and S. Kawaji on the negative magnetoresistance of two dimensional carriers in the MOS inversion layer observed at very low temperatures. They found that it was a phenomenon of orbital origin and not of spin origin, because only the perpendicular component of the external magnetic field on the interface was the important factor to determine the magnitude of this negative magnetoresistance. This results had already been interpreted by S. Hikami, A. I. Larkin and Y. Nagaoka within the context of microscopic theory of transport phenomena in random systems on the basis of newly developed theory of localization given by E. Abrahams, P. W. Anderson, D. C. Licciardello and T. V. Ramakrishnan in 1979. The results will open a new road to approach to the problem of localization and observed negative magnetoresistance effects even in usual three dimensional impurity band conduction. The other was the so-called quantum Hall effect. In the satellite conference of the 15th I.C.P.S. on high magnetic field at Hakone, K. von Klizing reported his precise experimental data of this effect and showed that the observed plateau value of Hall conductivity plotted against the gate voltage was given as $(e^2/h) \times$ (integer) so accurately that it might be promising to obtain the highest accuracy in an experimental determination of the Sommerfeld constant. The presence of the Plateau was very closely related to the presence of localized states in the tails of broadened Landau levels of the two dimensional system. The details of this interesting discovery will be discussed in this book by S. Kawaji and T. Ando. Thus one can find that the very fundamental physics involved in transport phenomena had been hidden in the most useful devices in modern electronics. One can also find an evidence for the fact that the development in technological and basic researches were closely related to each other, which is a traditional characteristic of semiconductor research.

Another example of a field which shows rapid development was the study of disordered or amorphous semiconductors. One can find plenary invited talks in almost very I.C.P.S. during the decade. They are:

"Electronic properties of amorphous semiconductors" by J. Stuke at Cambridge Conference 1970, "Current status of some basic problems in amorphous semiconductors" by W. Paul at Warsow Conference 1972, "Structures of amorphous tetrahedrally-bonded semiconductors" by G. A. N. Connell at Stuttgart Conference 1974, "Amorphous semiconductors" by N. F. Mott in Rome 1976 and "Defects in amorphous semiconductors" by H. Fritzsche at Kyoto Conference 1980. Although many papers were presented in every I.C.P.S. during the decade the way for reaching a systematic understanding of amorphous semiconductors seems to be very far away. A considerable progress has been made of microscopic mechanisms by hydrogenated amorphous silicon. Presumably the chemical view point will be very helpful to understand the complicated problems of amorphous structures. Further, amorphous states are metastable so that atomic rearrangements play an important role in thermal and optical processes. The phenomena of laser annealing discussed the 15th I.C.P.S. in Kyoto 1980 for the first time were examples to show the importance of such a rearrangement of atoms. The field was paid much attention from technological applications although the physics involved in the phenomena has not yet been thoroughly clarified.

The same view points are also very important for the interpretations of the phenomena observed in the electronic processes of deep impurity levels and in the atomic reconstruction of surfaces. Surface and interface studies from the microscopic viewpoint made remarkable progress during this decade as a result of new experimental methods such as LEED, RHEED, photo-emission, X.P.S., ESCA, use of synchrotron radiation and so on. Various structural and spectroscopic data on the atomic arrangement on the surface and its electronic states were accumulated in the latter half of the decade. The studies of surface reconstruction and of the presence of a long period superstructure such as 7×7 on the surface of Si and the shift of spectroscopic energies of surface atoms or ions detected by photoemission or XPS were examples of such data. In the field of deep impurity levels similar types of progress were recognized during the decade in the study of surface and interface. The rearrangement of host atoms at the interface or near the impurity atom is very important again as well as the dynamical displacement of these atoms. The correlation of trapped electrons plays also a very important role.

In the theoretical investigations of electronic structures in deep level impurities, in disordered or random systems and even on surfaces of atomic scale, one has to depart from the three dimensional k space representation and transform it into the R space representation which has been shown to be a powerful method for making connection with intuitive chemical bond pictures and for explaining various spectroscopic data. The term "bands and bonds" representing two different concepts were used during the decade. At the 10th I.C.P.S. in Cambridge 1970, J. C. Phillips gave his plenary invited talk on "Ionicity of the chemical bond in semiconductors". He also reported on the structure and properties of semiconductor surfaces in his plenary talk at the 13th I.C.P.S. in Rome 1976. In these talks he emphasized the importance of the chemical view point. M. L. Cohen gave his plenary talk at the 15th I.C.P.S. in Kyoto 1980, to present a firm theoretical basis to bridge these two concepts, "bands and bonds".

The relaxation of surrounding atoms and the correlation among trapped carriers are also very important in the theoretical studies of deep trapping states. The reconstruction of the surface atomic structures is in a simlar situation as I have already mentioned. G. A. Baraff and M. Schlüter presented a detailed theoretical calculation on the electronic structures of multiply charged vacancy in Si in which they showed the evidence of the so-called effective negative correlation energy as a cooperative effect of exchange, correlation and lattice relaxation. The investigation was reported at the 14th and the 15th I.C.P.S. in Edinburgh 1978 and in Kyoto 1980. In the previous decade G. Watkins et.al. made various interesting experiments on the electronic properties of a vacancy in Si mainly by the method of E.S.R. Now, a theoretical background was arranged for discussion of the observations done by G. Watkins et.al. quantitatively. As to the surface reconstruction, E. Tosatti was asked to give a plenary invited talk on "Semiconductor surface reconstructions" at the 13th I.C.P.S. in Rome 1976.

The other type of transformation is rather academic compared to the fields just mentioned. Strong coupling between carriers, or between carriers and phonons was a topic of this third decade. The strong coupling reveals various kinds of phase change.

A typical example is an electron-hole system. After Rogachev's talk in the 9th I.C.P.S. in Moscow 1968, the high density gas of electrons and holes became a very attractive system, because one can expect various phases, such as the exciton, exciton molecule or complex, electron-hole liquid or plasma. These phases are expected to be realized depending on the concentration of electrons and holes and on the temperature of the system. Controlled laser irradiation and indirect recombination processes produce favourable conditions to realize those various phases. "Condensation of excitons" was a title of a plenary talk at Warsaw Conference 1972 given by Ya Pokrovsky. E. Hanamura also discussed a similar problem, in particular its possibility of Bose-Einstein condensation from a theoretical point of view. His talks were found in the 10th and 11th I.C.P.S. in Cambridge 1970 and in Warsaw 1972. An experimental study to find the predicted condensation was reported by S. Shionoya at the 12th I.C.P.S. in Stuttgart 1974. This conference was remarkable in that many interesting papers were presented for electron-hole drops. Observation of microwave dimensional resonances of large electron-hole drops in Ge was reported by C. D. Jeffries, R. S. Markewicz and J. P. Wolfe. Luminescence of these drops was observed by G. A. Thomas, T. M. Rice and J. C. Hensel. Also a plenary talk on "Theory of electron-hole fluid" was given by T. M. Rice. From Japan, "The temperature and stress dependent stability of the drops in Ge" was presented by T. Ohyama, T. Sanada, F. Fujii and E. Otsuka, and luminescence properties of electron-hole liquid in highly excited pure and heavily doped Ge was reported by A. Nakamura and K. Morigaki.

The carrier-phonon interaction can also cause a phase change. At the 13th I.C.P.S., P. M. Pick gave a plenary invited talk on "Lattice dynamical aspects of phase transition". The IV-VI semiconductors such as PbTe and their alloys were studies by Japanese experimental physicists. At the 14th I.C.P.S. in Edinburgh 1978, papers on "Band gap and phase transformation in IV-VI semiconductors" and on "Observation of band structure change due to structural phase transitions in $Pb_{1-x}Sn_xTe$" were presented by K. Murase et.al. and K. L. I. Kobayashi et.al. Raman spectroscopy as well as high pressure measurement served as a powerful tool for such studies. The structural change caused by a change of carrier concentration was discussed by P. Kosin and N. Kristoffel.

Current interests in new phases of charge density wave, of Wigner lattice and of superconductivity promoted research for various new materials. Indeed, new material is a characteristic feature of this third decade. Layered compound of transition metal dichalcogenide, of graphite intercalation compounds, $(SN)_x$ and chained organic crystal such as TTF-TCNQ are typical examples of new materials. The research has been motivated not only by industrial needs to find materials for high T_c superconductors, for hydrogen storage or for batteries, but also by rather academic interests of many body interactions. In the 12th I.C.P.S. in Stuttgart 1974, A. P. Yoffe gave his plenary talk on "Electrons in two dimensional solids". In the 13th I.C.P.S. in Rome 1976, H. Kamimura gave his plenary talk on "One dimensional conductors" and F. J. DiSalvo read his invited paper on "Charge density wave instabilities in Layered compounds". "The density waves in one, two and three dimensions" was the title of a plenary talk by L. M. Falicov read at the 14th I.C.P.S. in Edinburgh 1978. In the 15th I.C.P.S., D. Jérome was invited to talk on the organic superconductor and he mentioned that a quasi one dimensional conductor $(TMTSF)_2PF_6$ was expected to show an interesting phase diagram suggesting the competing mechanisms of the so-called Peierls transition and superconducting transition.

T. Kasuya is a pioneer in the study of the magnetic semiconductors. In this decade, he reported on his study on magnetic polaron at the 10th I.C.P.S. in Cambridge 1970 and he was invited to give a plenary talk on "New aspects on the electronic properties of magnetic semiconductors". Spin dependent transport phenomena were also reported in a plenary talk by I. Solomon. In a plenary session at the 13th I.C.P.S. in Rome 1976, B. P. Zakharchenya was invited to talk on "Spin polarization induced by optical pumping in semiconductors". "A study of electron dynamics in CdS by spin flip scattering" was also reported by S. Geshwind, R. Romestain and G. Devlin at the same conference. A new category called "semimagnetic semiconductors" has been created in this field such as $Hg_{1-x}Mn_xTe$ and $Cd_{1-x}Mn_xTe$, having rather narrow band gaps and spin glass systems. Polish physicists showed much interest in this field. In the 14th I.C.P.S. of Edinburgh 1978, R. R. Galazka was invited to talk on these substances.

The other type of new materials is the superlattice. The search for new materials was not confined to a search for new compounds or crystals in the nature or by the chemical synthesis but also for physical methods to make artificial materials. We might say that one should design a new material. One of typical examples is the superlattice, a one dimensional artificial crystal by alternating stackings of layer of GaAs and GaAlAs obtained by the method of molecular beam epitaxy. Electronic properties of two dimensional carriers in the so-called modulation doped superlattice suggested a possibility for very high mobility of two dimensional carriers along the interface of each layer. The idea was soon applied to the GaAs-GaAlAs heterojunction to realize highspeed transistors. At the 14th I.C.P.S. in Edinburgh 1978, H. L. Störmer, R. Dingle, A. C. Gossard, W. Wiegmann and R. Logan gave a talk on the modulation doping. At the 15th I.C.P.S. in Kyoto 1980, H. L. Störmer was invited to talk on the high mobility by modulation doping of GaAs-GaAlAs heterostructure superlattice and interfaces.

The field of superlattice was initiated by L. Esaki, and he intended to study the tunneling current through stacking layers first. Now, the current along the interface was a very attractive subject analogous to that of MOS inversion layer.

So far, I emphasized the sign of transformation observed in the trends of research in this third decade. To summarize briefly, I will show Table 2. In the table, the transformations should proceed from left to right. The first three are those with respect to fields of study, the second three are with respect to phenomena or to theoretical bases and the last two are with respect to materials.

TABLE 2.

Sign of Transformations

Bulk ⟶ Surface and Interface
Shallow ⟶ Deep
Periodic ⟶ Random or Disordered

---

k Space ⟶ $\vec{R}$ Space
Weak Coupling ⟶ Strong Coupling
Static ⟶ Dynamic

Three Dimensional ⟶ Low Dimensional
Natural ⟶ Designed

Of course one can recognize steady and continual development in various traditional fields in semiconductor physics. For examples, band structure of Tellurium was studied both theoretically and experimentally so that the nature of camel-back structure of valence band near k=0 point could be clarified.

The talks given by M. Hulin, by G. Dresselhaus and M. S. Dresselhaus, and by H. Kamimura, K. Nakao and T. Doi in the 10th I.C.P.S. in Cambridge 1970 were examples. These studies stimulated the investigation of Te very much and the impurity states and optical rotary power were studied by K. Natori et.al. theoretically. As to the complicated Fermi surface of graphite, an important experimental result was given by P. R. Schroeder, M. S. Dresselhaus and A. Javan in the latter half of the previous decade. It was established experimentally that the carriers near k=0 was holes.

Band calculations of graphite were done many theoretical solid state physicists because of typical layer structure and of consisting of simple atoms of carbon. However, no success was obtained to explain the presence of holes near k=0. K. Nakao, M. Tsukada, H. Nagayoshi and Y. Uemura studied the band calculation of graphite more thoroughly by taking into account the interaction between different layers properly and finally they obtained the result to be able to realize the observed Fermi surface theoretically in detail as well as many structures observed in the optical spectra of graphite. The final report was presented at the 13th I.C.P.S. in Rome 1976 by H. Nagayaoshi and Y. Uemura.

In the field of transport studies, the investigations of impurity conduction and of the hot electron effect were showing continual development. W. Sasaki and his collaborators were attracted to the field since the latter half of the first decade. In this decade they developed the studies on NMR investigations in heavily doped Si, on specific heat and magnetic properties of Si heavily doped by P and on the variable range hopping conduction in the lowest temperature range. W. Sasaki was asked to give plenary talk on "Impurity conduction in Ge and Si" at the 15th I.C.P.S. in Kyoto 1980.

The hot electron field was also very much developed during the decade. The name, "hot electrons" has one of the longest life times as a title of sessions in every I.C.P.S. During the decade, very specific distributions of carriers in phase space caused by high fields such as those caused by the so-called typical streaming effect were studied. The effect was found in the high resistivity polar semiconductors such as AgBr. S. Komiyama, T. Masumi and K. Kajita reported on the anomalous distribution of hot electrons in silver halides in crossed electric and magnetic fields at the 13th I.C.P.S. in Rome 1976. Furthermore, at the 14th I.C.P.S. in Edinburgh 1978, S. Komiyama, T. Masumi and T. Kurosawa presented their paper on "A new mode of streaming cyclotron motion of hot electrons at intense microwave fields in AgBr". The distribution of carriers discussed in these reports was very much different from that originally obtained in therms of hot electron temperature.

I have referred mainly to the Japanese contributions, here, as examples of continual development. It is quite sure that many other important contributions classified in this category were presented at every I.C.P.S. from many countries.

To sketch the activities in this decade, I did not divide my description into sub-topics such as international and domestic, or academic and industrial, as done by the descriptions in §2 and §3. This is of course a matter of convenience. However, it is also true that to divide the descriptions into two, international and domestic, has lost its meaning in this decade. As shown in Fig. 1, the number of accepted papers by Japanese authors was increasing exponentially from the second decade so that there seems to be no reason to discuss the trend of domestic activities separately from the international one.

5. Closing Section

It was our great pleasure that we had the 15th I.C.P.S. in Kyoto again in 1980, fourteen years after the first Kyoto Conference in 1966. The chairman of the second Kyoto Conference was Prof. H. Kawamura and its general secretary was Prof. H. Kamimura. I was asked to present the closing address for the second Kyoto Conference as its vice-chairman. The description of the previous section are mainly based on this closing address. To close this article, I will directly refer to the final part of this closing address again for the closing of this article.

What about the next decade? I do not know. Instead of playing the

TABLE 3.

There is no other body with which I am acquainted, that, like sulphuret of silver, can compare with metals in conducting power for electricity of low tension when hot, but which, unlike them, during cooling, loses in power, whilst they, on the contrary, gain. Probably, however, many others may, when sought for, be found.

role of a fortune teller, I will call the reader's attention to a "new material" discussed nearly 150 years ago. During the course of his studies on electricity in various materials, the famous Mickael Faraday was surprised when he discovered the strange temperature dependence of electricity of sulphuret of silver. It was indeed a new material at that time in the terminology of I.C.P.S. After describing his careful observations of the interesting characteristics of this material when it was heated on the lump, he closed his detailed descriptions by the sentences shown in Table 3 which were published in the Philosophical Transactions in 1833. We may choose this date as the birthday of our semiconductor physics.

Readers can sense from these sentences, Faraday's almost naive and simple surprise and joy over his discovery. You can also sense his optimism that many new materials will be found in the future. His optimism was not brash or contrived to get more research fund from the government or a foundation; it was based on his conviction of what nature held in store for the future generations of scientists. If we all follow his spirit of optimism, I think that the progress of semiconductor research in the following decade will be as great as that in the past and that many of the readers of this book, particularly younger scientists will experience many surprises and joys as experienced by Faraday.

Finally, I ask the reader's indulgence because I am afraid that my sketch of the development of semiconductor physics covering these thirty years may be too personal so that I have committed sins of omission and commission when discussing each activity.

Lastly, I will express my sincere thanks for warm-hearted friendship of many friends in the field of semiconductor physics not only of the friends in Japan but also of the friends in many foreign countries. It was quite certain that I could not have spent a pleasant period in these thirty years without their friendship.

Special thanks are due to the editors, Prof. Y. Toyozawa and Prof. H. Kamimura to arrange a symposium in my honor at the occasion of my retirement from the University of Tokyo, and due to Prof. S. Yamamoto for his kind advice on the English composition of this article.

References

1) Y. Uemura: "Closing address" Proc. 15th Int. Conf. Physics of Semiconductors, Kyoto, 1980, J. Phys. Soc. Japan 49 (1980) Suppl.A, p.1309. The descriptions in §4 and §5 are based on this article.

2) Y. Uemura: "A quarter of century for the semiconductor research — Japanese semiconductor physics after the World War II. — "Buturi" Vol.32, No.10 (1977) p.10-27, in Japanese. The descriptions in §1, §2 and §3 are based on this article.